An Ethics of Place

An Ethics of Place

Radical Ecology, Postmodernity, and Social Theory

Mick Smith

State University of New York Press

Published by
State University of New York Press, Albany

Printed in the United States of America

For information, address State University of New York Press,
90 State Street, Suite 700, Albany, NY, 12207

Production by Marilyn P. Semerad
Marketing by Fran Keneston

Library of Congress Cataloging-in-Publication Data

Smith, Mick, 1961–
An ethics of place : radical ecology, postmodernity, and social theory / Mick Smith.
p. cm.
Includes bibliographical references and index.
ISBN 0-7914-4907-6 (alk. paper)—ISBN 0-7914-4908-4 (pbk. : alk. paper)
1. Deep ecology—Moral and ethical aspects. 2. Environmental ethics. I. Title.
GE195 .S54 2001
179'.1—dc21

00-041298

10 9 8 7 6 5 4 3 2 1

This book is dedicated to Jim Davidson

1965–2000

Contents

Acknowledgments

This book would never have been completed without the help of innumerable people. Above all I would like to thank Joyce Davidson for her thoughts and conversation on many, if not all, of the issues this book seeks to address. I am especially indebted for her contribution to the original joint conference paper that forms the basis for chapter 7 and her comments on the final version of this chapter. I would also like to thank colleagues at the University of Abertay Dundee who read and commented on draft versions of various chapters, Alex Law, Hazel Work, Andy Panay, and Jason Annetts, and to the two anonymous readers who made many valuable suggestions. I owe a special debt to Andrew Brennan for his encouragement and support over the years and I'd also like to thank Arthur Callaghan for bearing with my earlier philosophical speculations when I was supposed to be studying soil fungi. Thanks also to Val Plumwood, Richard Twine, Louise Scott, Jim Valentine, John Stewart Watson and to Zina M. Lawrence and Dale Cotton, both formerly of SUNY Press and Marilyn Semerad, Production Manager of SUNY Press. As usual, none of the above are in any way or form to blame for all my errors that remain as yet undetected. I would also like to thank three very special nonhumans who were present during much of its writing, Mork, Rionach, and Sammy.

I am very grateful to the following for granting permissions to incorporate copyright material that has previously been published elsewhere. Blackwell Publishers of Oxford for "Letting in the Jungle," *Journal of Applied Philosophy* 8.2 (1991): 145–54. John Wiley and Sons Ltd. of Chichester for "A Green Thought in a Green Shade: A Critique of the Rationalization of Environmental Values," in Yvonne Guerrier, Nicholas Alexander, Jonathan Chase, and Martin O'Brian (Eds.), *Values and the Environment: A Social Science Perspective* (1995): 51–60, copyright John Wiley & Sons Limited, reproduced with permission. The Center for Environmental Ethics at the University of North Texas, Denton, for "Cheney and the Myth of Postmodernism," *Environmental Ethics* 15.2 (1993): 3–17, and "Against the Enclosure of the Ethical Commons:

Radical Environmentalism as an Ethics of Place," *Environmental Ethics* 19.4 (1997): 339–53 and "To Speak of Trees: Social Constructivism, Environmental Values, and the Futures of Deep Ecology," *Environmental Ethics* 21.4 (1999): 359–76. "Environmental Antinomianism: The Moral World Turned Upside Down?" is reprinted from *Ethics and the Environment* 5.1, copyright 2000 with permission from Elsevier Science of Oxford. Special thanks for arranging this last permission to Helen Wilson of Elsevier and Victoria Davion.

Last, but certainly not least, I would like to thank my mother and father, Eileen and Alan Smith, for all their support and love over the years.

Introduction

Mean while the Mind, from pleasures less,
Withdraws into its happiness:
The Mind, that Ocean where each kind
Does streight its own resemblance find;
Yet it creates, transcending these,
Far other Worlds, and other Seas;
Annihilating all that's made
To a green Thought in a green Shade.
—Andrew Marvell, 'The Garden'

The ability of thought to transcend the circumstances in which it finds itself, its urge to create "other worlds," is surely one source of our present environmental problems. It is also the wellspring of a hope that we might overcome such problems.

Modernity, the social condition of our contemporary world, is characterized in thought and in deed, by its Promethean striving to go beyond all given limits. This continual and accelerating movement is the basis for that dominant myth of progress that typifies and justifies the modernist enterprise. Through increasing the speed and efficiency of our productive labors, modernism dreams of defying Earth's gravity, of achieving an "escape velocity" that would take us beyond all natural constraints. Of course, though the pace of life and invention increases daily, we have not yet succeeded in taking flight from our planetary imprisonment. Even the point at which we might so liberate ourselves is not quite in sight; it seemingly always lies just beyond the next horizon. But can anyone seriously doubt that we are indeed making progress, that we are well on our way to taking control of our destiny?

Yet doubts persist. The problem is that the higher our ambitions fly the thinner the air becomes. As the silent pull of nature's gravity is loosened, so we increasingly find ourselves dependent upon artifice to survive in an atmosphere that is, if anything, more hostile to our mental and physical well-being than that we sought to leave behind. For despite its promise, the vacuum of abstract space is not conducive to either thought or life, both of which require the oxygen that only contact with the oceans, rivers, and woods of this world can supply.

The desire to "encounter strange new worlds" is all very well; but this constant search for novelty is achieved at the cost of erasing those places and their inhabitants on this world that we, caught up in our own concerns, have come to regard as strange(rs). For in launching ourselves forward into the as yet unknown, we forget that every action we take has innumerable reactions, reactions that are not necessarily equal and opposite, and for that reason not predictable. And, forgetting this, we continue to be tipped off balance as our best-laid plans fall through. When this happens, we plummet down through fume-laden skies toward a landscape increasingly scarred with the marks of past attempts and failures to rise above it.

And what if, through the application of technology and the triumph of technological rationality, we should eventually succeed in our ambition to be as gods? What if, at some future point on our journey to these heavens of our own invention, we should find the time to rest from our labors and, looking back, seek to survey our achievements. Will they seem good to us?

To many this seems unlikely. Environmentalists are among those who stress that in modernism's mad rush we have often lost as much or more than we have gained. The instruments of environmental destruction, the chainsaw and the car, capitalism and simple carelessness, have cut and gouged and burnt their way through the planet's delicate fabric. Species have been wiped from the face of the globe, forests felled, and seas drained. The oily glint of shining new machinery comes at the cost of the oiled feathers of seabirds and the paper profits of the companies concerned are brought at the cost of the earth itself.

But, despite the despair that is an inevitable consequence of having a ringside seat at the ruination of a beautiful and invaluable world, this environmental holocaust also breeds resistance. This resistance takes many forms; the Chipko women in India, Earth First! in North America, the antiroads protests in Britain, and many, many more. These radical protests are but the tip of an environmental movement trying to halt deforestation and save the seas. Environmentalists are joining with others who find their concerns cast aside or crushed by modernism's insatiable desire to transform all within its grasp. This combined resistance emphasizes thought's creative rather than destructive potential, its capacity to think beyond the narrow confines of modernism's own concerns. It re-envisages a future where we walk lightly on the Earth, feeling gravity's caress and breathing deeply of its sweet airs rather than spiraling like crazed satellites in the void above. What follows hopes to take *one small step* away from that which would annihilate all that's made, and toward a green thought in a green shade.

Introduction

In his theses on the philosophy of history Walter Benjamin describes a

> painting named "*Angelus Novus*" [which] shows an angel looking as though he is about to move away from something he is fixedly contemplating. His eyes are staring, his mouth is open, his wings are spread. This is how one pictures the angel of history. His face is turned towards the past. Where we perceive a chain of events, he sees one single catastrophe which keeps piling wreckage upon wreckage and hurls it in front of his feet. The angel would like to stay, awaken the dead, and make whole what has been smashed. But a storm is blowing from Paradise; it has got caught in his wings with such violence that the angel can no longer close them. This storm irresistibly propels him into the future to which his back is turned, while the pile of debris before him grows skyward. This storm is what we call progress. (Benjamin, 1992: 249)

Radical ecology is an ethical and political protest against this seemingly irresistible hurricane of destruction, a protest against a mode of existence that has been largely insensitive to the environmental devastation "progress" has left in its wake. It is a protest that recognizes that the problems of deforestation, ozone depletion, urban smog, loss of biodiversity, and climate change, cannot be treated in isolation. They stem from, and are entwined with, our modern forms of life. For this reason the critique of environmental destruction necessarily becomes a critique of contemporary society.

The premise of this work is that, for the most part, theorists in philosophy and the social sciences have failed to even recognize, let alone support, the serious ethical challenges that radical environmentalism presents to both our contemporary way of life and their accepted *modus operandi*. For the most part they have been happy to squeeze environmentalism into current paradigms and debates, treating it as an academic resource capable of providing novel exemplars for tired arguments or revitalizing flagging careers. Even where environmentalism has been taken seriously, for example, as a challenge to political theories like Marxism or liberalism, it has found itself either parodied to such an extent that it becomes almost unrecognizable, or eviscerated and then absorbed in more palatable form into frameworks with which it has little in common.

The problem I have set myself here is one of developing a theoretical account that remains true to the spirit of radical ecology and yet is capable of engaging with modernity's established problematics. This does not, I hope, entail compromise, but it is nonetheless an attempt to communicate and recognize commonalties with other perspectives. For, however hard it tries to break free, to imagine a different order of things, radical environmentalism is inevitably caught up in current social practices and worldviews. Even in their most utopian forms radical environmental protests are largely immanent critiques, they arise from and express contradictions within the society they seek to criticize. As such they cannot hope to, and usually do not want to, escape *all* contemporary influences.[1] Having said this, environmentalism offers a vigorous and wide-ranging critique of many aspects of modernity, its institutions and ideologies, its economics and ethics, its culture and its creeds.

Modernity and Postmodernity

All this, of course, raises questions about the extent to which a radical ecological critique might be considered "post-modern," questions that Jim Cheney's (1989) seminal essay "Postmodern Environmental Ethics: Ethics as Bioregional Narrative" sought to address. Max Oelschlaeger describes Cheney's article as "the most catalytic essay in postmodern environmental ethics" (Oelschlaeger, 1995: 10). Although many others have now taken up this issue (e.g., Zimmerman, 1994; Gare, 1995; Conley, 1997; Esteva & Prakesh, 1998), Cheney's essay continues to provide a benchmark for understanding the relationship between modernism, postmodernism, and ecological ethics, not least through its emphasis on context, or *place* in determining ethical values. However, Cheney posits an absolute dichotomy, between a deluded modernism characterized as producing foundationalist, essentialist, colonizing, and totalizing narratives and a contextual postmodernism; a dichotomy that is problematic in more than one respect (Smith, 1993).

Cheney's essay centers on an explicit thesis that divides human history into three epochs: premodern, modern, and postmodern. According to Cheney, the change from the premodern to the modern period began "some nine or so millennia" ago with the appearance of agriculture and the change from modern to postmodern only very recently. The dominant worldview of the whole of Western society and consequently all Western philosophy until now has been modern.[2] (Cheney thus differs markedly from those philosophical and sociological conceptions of modernity that regard Descartes as the first truly modern philosopher or associate modernity with the social changes based on industrialization, the growth of science and technology, the nation-state, capitalism and so on [Featherstone, 1988].) For Cheney, postmodernism heralds a long-awaited return to a "primitive" understanding of our place in the world. "With the advent of postmodernism, contextualized discourse seems to emerge as our mother tongue; totalizing, essentializing language emerges as the voice of the constructed subjective self, the voice of disassociated Gnostic alienation" (Cheney, 1989: 122).

Cheney presents a myth of a past Golden Age with a tribal humanity at one with itself and nature, a present alienated society of atomistic individuals divorced from the natural world, and a future postmodern utopia, each stage epitomized by the form of language it uses.[3] According to Cheney, the premodern paradigms of contextual discourse were tribal mythologies. Myths are both fabulous stories with moral connotations and forms of "knowledge shaped by transformative intent" (Cheney, 1989: 121). They are "historically sociologically and geographically shaped system[s] of reference that allow . . . us to order and thus comprehend perception and knowledge" (Paula Gunn Allen in Cheney, 1989: 121). Mythic narratives emerge from a "meditative openness to the world" that allows the natural world to "speak[s] through us" (Cheney, 1989: 119). Tribal societies' residence in natural contexts ensures the contextuality of their discourse; "the world discloses itself by our being *rooted* in the world" (119).

By contrast, fallen (modernist) language "uproots itself"; it is a form of theory abstracted from place and context that claims to provide an acontextual and universal description of the way the world operates. Once divorced from the specific settings and practices that originally gave it meaning, "language closes in on itself, becoming inbred" (126). For Cheney, the totalizing discourse of modernism is indelibly associated with the artificial, the unnatural, and the colonizing, that which is abstracted and applied outside its own remit—a remit that in natural circumstances is bounded within a biogeographical region. "The possibility of totalizing, colonizing discourse arises from the fact that concepts and theories can be abstracted from their paradigm settings and applied elsewhere" (Cheney, 1989: 126). By contrast it is simply not possible for mythic discourses to be applied out of place: "[T]hey are not thought of as exportable" (120). To some extent this makes sense, since it is easy to imagine how a colonizing language, developed in surroundings very different to its new circumstances, *might* disrupt the delicate balance that Cheney thinks was reached between "primitive" humanity and its natural environment. Just as an introduced species may not be at home in a temperate climate and either does not flourish there or destroys indigenous flora and fauna, so taxonomies designed for one place *might* be disruptive in other places.[4]

The postmodern solution then seems obvious—in order to find our roots and evade the environmentally destructive effects of modernity's all pervasive discourse we must (re)turn to nature. "Is there any setting, any landscape, in which contextualizing discourse is not constantly in danger of falling prey to the distortions of essentializing, totalizing discourse?

Perhaps not. A partial way out might be envisioned, however, if we expand the notion of a contextualizing narrative of place so as to include nature—nature as one more player in the construction of community" (128). In other words Cheney emphasizes the import of natural environments in providing an anchor for our narratives and our values. "Bioregions provide a way of grounding narrative" (128).

I hope that it will become obvious as this work unfolds that I wholeheartedly share Cheney's view that nature has to be recognized as playing an active part in the construction of those narratives, values, and communities that might re-place certain aspects of modernity. This will entail, as he suggests, that we come to understand "[o]ur position, our *location,* . . . in the elaboration of relations in a non-essentialising narrative achieved through a grounding in the geography of our lives" (126). However, there remain important differences between our respective positions, and it is also necessary to be clear about the senses in which each of our theses might and might not be regarded as postmodern.

First, though Cheney recognizes that some tribal peoples fail to realize their social or environmental potential (in terms of conforming to his ecological expectations) he consistently romanticizes their lives and their relationships to their natural environs. But, even if one accepted his contentious claims about the biogeographic contextuality of prehistoric languages, there is no reason to suppose that such languages have been uniformly environmentally friendly. Environments and tribal peoples have never been static and the development of language and place may have been one that saw the destruction of many features of the original prehuman landscape.[5] Many species seem to have been victims of the expansion and movement of early hunter-gatherer peoples. The giant Moa, an ostrichlike bird, was extinct within a few hundred years of the arrival of the first people in New Zealand. Java lost its pygmy elephants, Madagascar its pygmy hippopotamus, and the destruction was even more marked in North America where two-thirds of the mammalian megafauna became extinct between 10,000 and 12,000 years ago (i.e., in Cheney's premodern period) following the influx of people across the Bering Strait.[6] An examination of contemporary hunter-gatherer societies similarly suggests that it is impossible to generalize about their impact on and attitudes toward the environment. As the environmental historian and geographer I. G. Simmonds, states "the picture is one of variability and contingency. . . . However, any picture of hunter-gatherers simply as responsive children of nature living solely off a provident usufruct is part of a myth of a Golden Age" (Simmonds, 1994: 8).[7]

Such Arcadian romanticism can itself be traced back to the colonial expansion of European powers and the changes in economic and social structure that preceded and accompanied the advent of modernity (in its more usual sociological sense). The sixteenth and seventeenth centuries saw the beginning of a sea change in attitudes toward nature that was later to culminate in Romanticism's ethical critique of the mechanistic worldview of the sciences and the negative impacts associated with industrialization (Thomas, 1984; Merchant, 1990). Europe's "discovery" of indigenous peoples untouched by Western civilization yet seemingly leading idyllic lives was an important catalyst for these attitudinal changes (Grove, 1995).

Jean-Jacques Rousseau, a central figure in Enlightenment thought, is a case in point. Like Cheney, Rousseau was inspired by an Arcadian myth of a prehistoric society rooted in nature. Rousseau's depiction of an Edenic state of nature peopled by "noble savages" was influenced by the findings of contemporary explorers like Bougainville, whose reports of Tahiti and its populace made it appear paradisiacal.[8] Unfortunately, Bougainville actually had only the most superficial acquaintance with the customs of the islanders and Tahitians' social and natural relations were not all that the Romantic might have envisaged. That Tahitian society had strong "class" divisions and was adept at human sacrifice were facts that emerged as contacts with the culture became more prolonged. At the time of Tahiti's "discovery" by Europeans it had a massive population of approximately 200,000: "A single bread-fruit tree was often owned by two or more families, who disputed each others' rights of property over the branches. Infanticide was habitual" (Adam, 1976: 6). Douglas C. Oliver claims that violence was endemic and mass rape of those females on the losing side in battle a common occurrence. "Sometimes when a warrior felled his opponent he would beat the body to a flat pulp, cut a slit through it large enough for his own head to pass through, and then wear it, poncho fashion as a triumphant taunt" (Oliver, 1974).

This is not to say that modern society is morally superior to ancient Tahitian society (as the scale and horror of recent wars attest). Nor is it to claim that modernists have nothing to learn from primary peoples (or, for that matter from Romanticism). Quite the contrary. However, we do no one any favors by adopting a framework that fails to recognize the cultural and moral complexities of other societies and their environmental relations. More importantly we have to be reflexively aware of the origins and operation of the complex and often conflicting moral narratives to be found within our own modern society (of which Romanticism is one). Modernity and modernism need to be seen in the context of their specific historical and cultural settings.

Cheney admits that his essay has not yet "begun the work of [social] negotiation for the culture in which [he] finds [himself]" (Cheney,

1989: 134). But to do so we must recognize both that modernity has its own myths, influenced by its own interactions with particular geographical backgrounds, and that there is no *a priori* reason why even extreme modernist narratives do not deserve attention as examples of a world speaking through people. This world may largely be one of artifice but it is nonetheless a part of nature in the wider sense in which humanity is natural too. Cheney is willing to extend this contextual privilege to some instances of modernist discourse where the predominant influence is supposedly natural, for example Aldo Leopold's land ethic, but can't other modernist claims be seen as analogously *rooted*?[9] Can't they too be seen as instances of particular places operating, in Cheney's words, "all the way up"? In a different context Cheney has asked a similar question: "*Might* we listen with the same ear to the residents of Harlem (or to the corporate executives engaged in the destruction of the old-growth forests) with which we listen to the voices of a tall-grass prairie in southern Wisconsin? What would such a listening be like?" He seems to think that we might and must achieve this listening. "Even the strategies of the colonizers must be understood ecologically. They are not to be understood or condemned using timeless and ahistorically 'true' criteria" (Cheney, 1989a: 317–18).

But if we are to listen attentively to modernity's multifarious voices then we have to recognize that the residents of Harlem and the corporate executives hell-bent on trashing old-growth forests occupy very different places within it.[10] To develop, as Cheney wishes, a "truly revolutionary practice" that does not disenfranchise oppressed people (with the possible exception of tribal peoples) we must recognize the cracks, fissures, and tensions within modernity itself. We can't treat modernity, as Cheney does, as a unified whole finding expression in a closed and alienated narrative with a single way of speaking of and experiencing the world. Nor can we depend upon an idealized picture of a timeless natural environment, as opposed to the historicized artificial environment of our cities, to escape the necessity for a constant recontextualizing of social discourse. To do so is in its own way to accept a nature/culture division that is undoubtedly modernist in origin.

Things are not as simple as Cheney's neat epochal distinctions suggest; the boundaries Cheney draws between different narrative regimes and their respective cultures are not clear cut. His own, presumably postmodern, discourse is a case in point, since according to his own criteria it has all the hallmarks of a modern abstract theory. It is *foundationalist* insofar as it makes bioregions the necessary grounds for all properly contextual discourse.[11] It is *colonizing* to the extent that it appears in an international journal written in English, the most widespread colonial language. It is *essentializing* in its conception of all modernism as inherently divorced from place and *totalizing* in its pretension to provide a universal human history.

Of course, such incongruity might simply be taken as an indication of the tenacity of modernity's grasp on our thinking. Perhaps, despite his best postmodern efforts, Cheney has fallen unwilling victim to a subliminal return of an inadequately repressed desire to essentialize and totalize. On the other hand, the fault may lie with some of Cheney's criteria that seem equally helpless when it comes to distinguishing between primitive and modern narratives. Actual evidence of whether the discourses of premodern societies were contextual as Cheney claims or totalizing is difficult to come by, for by definition "prehistoric societies leave no discourse for posterity. Even if we can, as Cheney claims, take the discourse of contemporary (so-called) primary peoples as exemplifying those of prehistoric cultures the evidence remains inconclusive.[12] But most importantly, Cheney obscures what is at issue by conflating contextuality, in the sense of the particular environmental context within which a discourse is produced, with the contextuality of that discourse's claims, that is, whether or not it makes *universal* claims. He provides no substantial evidence to support a link between bioregional contextuality and discursive contextuality in terms of bioregions harboring antifoundationalism and antiessentialism, or leading to the rejection of linguistic colonialism or totalizing narratives. He certainly cannot hold that tribal peoples do not make totalizing generalizations. Indeed, in introducing the example he offers of a contextual discourse, that of the Ainu, the indigenous people of Japan, Cheney actually claims that "*everything* is a Kamui [spirit] for the Ainu" (Cheney, 1989: 128, my emphasis). If this is true then it certainly seems to suggest that while the Ainu may be contextual in the sense that they come from a particular locale and have a language influenced by that locale's natural context, they are not averse to making universal and totalizing generalizations about the world.

Similarly, it simply isn't plausible to divide the modern from the postmodern on the basis of the latter's contextuality, if this is somehow supposed to mean both its avoidance of abstraction and its (potential) rootedness in bioregional place. While Cheney is silent about the origins of postmodernism's often extremely esoteric narratives, almost all commentators emphasize postmodernism's links with urban environments. For example, Charles Jenks claims that one of the architectural canons of postmodernism is to produce an "*urbane urbanism*. . . . Urban contextualism gains near universal assent. New buildings, according to this doctrine, should both fit into and extend the urban context" (Jenks in Docherty, 1993: 285).[13] More than this, the antifoundationalism typical of postmodernism is often regarded as expressing (and in some cases supposedly celebrating) rootlessness in a world dominated by the constantly accelerating circulation of disconnected values, symbols, narratives, and so on. Thus Jean Baudrillard regards our hypermodern world where, "the only kind of profound pleasure is that of keeping on the move" (Baudrillard, 1994: 53), as entailing a less, not more, immediate relation to nature since nature and culture become subsumed in a new, abstract system of manipulable signs, an infinitely flexible currency of exchange without any grounds whatsoever.

Clearly Cheney is operating with a notion of postmodernism with only tenuous connections to that of either Jenks or Baudrillard.[14] However, my point is not just one about the diversity of opinions over/within postmodernism. Rather, I am arguing that, if we accept the argument that the form and content of narratives are linked to their time and place, this means that postmodernism is itself best understood in terms of its origins within a modern, urban environment. The arrival of postmodernism is not, in itself, an indication of the end of modernity. Rather, as a recent product of a society that is arguably the most extreme and expansive form of modernity, postmodernism is itself both riven by and responding to modernity's contradictions.

This is why, while offering several possible definitions of postmodernism, Jean-François Lyotard argues against regarding postmodernism as a clean break with past (modern) traditions where "the 'post-' of postmodernism has the sense of a simple succession, a diachronic sequence of periods in which each one is clearly identifiable. . . . [This] idea of a linear chronology is itself perfectly 'modern'" (Lyotard in Docherty, 1993: 47–48). Postmodernism "is undoubtedly a part of the modern" (44) though not in the sense of a continuation or exacerbation of past trends. For Lyotard, "the 'post-' of postmodern does not signify . . . a movement of repetition but a procedure in 'ana-': a procedure of analysis, anamnesis, anagogy, and anamorphosis that elaborates an initial forgetting" (50). In other words, postmodernism is reflexive about and recalls those aspects of modernity that modernism would rather forget.[15]

We need then to distinguish between *modernity* as an epoch or a social formation with certain salient and distinguishing features and *modernism* as the dominant but not sole ideology/narrative within that society.[16] Just how we characterize modernity and modernism will be the subject of much that is to follow. Suffice it to say here that, while modernism as a narrative has tended to support and even exacerbate the features of modern society that cause most consternation to radical ecologists, other countermodern discourses have always existed within modernity, as the case of Romanticism shows.[17] Some postmodern discourses undoubtedly add to this countermodern critique, others are more clearly tied to, and even seek to exacerbate, modernity's current trajectory and as such might themselves be regarded as forms of hypermodernism. This means that we cannot look to postmodernism per se for our salvation; equally we cannot reject it out of hand since certain elements of postmodernism might ally themselves with an environmental critique of contemporary, modern society.[18] This ambiguity is recognized by Frederick Jameson, who reluctantly adopts the term "postmodernism" as a label for the cultural logic of late capitalism; "for good or ill we cannot *not* use it" though it is "internally conflicted and contradictory" (Jameson, 1991: xxii).

Most postmodern discourses are, like their modernist predecessors, neither wholly critical nor wholly compliant with the current social order. As such, we need, quite literally, to understand where they are coming from and they need to be deployed tactically and reflexively in environmental critiques. Lyotard's own work is a case in point. Cheney's skepticism of totalizing narratives obviously draws upon what Best and Kellner characterize as Lyotard's advocacy of "'incredulity toward metanarratives,' the rejection of metaphysical philosophy, philosophies of history, and any form of totalizing thought. [This is because the] metanarratives of modernity tend, Lyotard claims, towards exclusion and a desire for metaprescriptions" (Best & Kellner, 1991: 165–66). This corroborates Cheney's usage. However, Lyotard would find difficulty in turning to bioregional diversity as a guarantor of plural but naturally bounded narratives. There can be no return to what Lyotard regards as self-enclosed communities where "[n]arrative is authority itself. It authorizes an unbreakable *we,* outside of which there can only be a *they*" (Lyotard in Gare, 1995: 65).[19] Far from being motivated by nostalgia for the real (nature), Lyotard regards postmoderns as "witnesses to the unpresentable" (Lyotard, 1991: 82). Against the totalitarian imposition of powerful unitary narratives he posits a micropolitics of desire and a multivocal pluralism of different language games within as well as between communities. His preferred strategy is to wage "a guerilla war" of constant subversive interventions in order to undermine the authority of any discourse that threatens to attain hegemony. The best hope for achieving this depends, he thinks, upon the extension of free public access to computerized information systems and the consequent free play of language itself (Lyotard, 1991: 67).

Lyotard highlights problems stemming from the tendency for narratives to become authoritative and authoritarian, problems that exist in their own ways in both premodern and modern societies, but he offers no easy way out. His philosophy is itself locked into a contest with(in) modernity since it is "the crisis of modernity, which is the state of postmodern thought" (Lyotard, 1997: 101).[20] Though not without its critical and utopian aspects,[21] Lyotard's work could not be used straightforwardly to condemn modernity or condone a return to a romanticized prelapsarian harmony with nature since postmodernism is, Lyotard claims "a fable that signifies the end of hopes" (Lyotard, 1997: 100).

In other words, we must recognize that postmodernism has a complex genealogy prefigured by discourses in that modern society within which it too originates. We have to be wary of amalgamating, as Cheney does, all past Western philosophical traditions, irrespective of their disparate backgrounds and complex interrelationships, under the single heading, *modern*. In conflating modernity (the epoch) and modernism (a particular ideology/narrative) Cheney effectively disenfranchises modernity's inhabitants who have, over nine millennia, apparently all led inauthentic and alienated existences, separated from the natural context that might have endowed them with an authentic mythic voice. But Cheney never explains why city, town, and agricultural landscapes, occupied by the vast majority of people, past and present, cannot also be places that speak through us in a manner analogous to the way that the 'natural' world is supposed to speak through premodern peoples. To be sure, the city environment is not one populated with salmon, unless they lie cold on the supermarket slab, but it is populated with its own ecology of cars preying upon pedestrians, the rich upon the poor, of smog, pollution and resistance. These become part of the narratives of our lives, of our urban and suburban myths.[22]

Cheney's historical divisions also seem to disenfranchise tribal peoples by relegating them to little more than vectors functioning to give voice to those natural environs that speak through them. But, to quote Hans Peter Duerr, "[i]t seems a mistake on which extreme relativists and dogmatists of the 'transcendental' bent agree . . . that we do not think the myths but the myths think themselves in us" (Duerr, 1985: 97). All people (individuals) and all languages are relatively autonomous whatever their bioregional limitations. All peoples (including tribal peoples) are able to use language to stand back from as well as to express their place of origin. Depending on which aspect you want to emphasize, it is equally true (and equally false) to say that we are all free to produce our own myths or alternatively we are all tied to producing the myths of our place.[23] It is precisely in this space, between freedom and constraint, (whether individual, linguistic, material, etc.) that thinking, caring, and critique maintain their tenuous existence.

In summary, a number of things flow from this analysis. While we have good reasons for linking environmental problems with the crisis of modernity we cannot regard postmodernism as representing the "Second Coming" of a narratival innocence. Postmodernism is by no means an "Immaculate Conception," free from the taint of modernity's original sins, nor, despite Cheney's born-again advocacy, does it offer us redemption. Postmodernism cannot do these things because it too has to be understood in terms of its social origins. It does not occupy or express a unified position but reflects, in its forms and contents, as modernity's own discourses do, the fragmentation of contemporary society. Postmodernism is the illegitimate child that straight-laced modernists refuse to recognize as their own. In finding itself expelled and homeless it seeks an explanation for this (postmodern) condition, through disputing both the legitimacy of the current social order and the degree of its inherited "family resemblance." Thus, while the marginal position of postmodern theories enables them to offer much needed insights into the "dark heart" that often lies concealed beneath modernity's superficial civility, they are by no means alone in doing so. Modernity has been marked by what amounts to an obsession with self-understanding, an obsession epitomized by social theory. To ignore, as Cheney does, the claims of discourses—whose entire purpose has been to illuminate modernity's recesses—just because they are the recognized offspring of their own modern times, may prove suicidal.

This is why this work seeks to reconstitute (rather than simply apply, like a sticking plaster) a theoretical understanding of the environmental crisis and of ethics through an immanent critique of social theory and philosophy that calls upon modern and postmodern perspectives. While recognizing the manner in which every theory is a mythic expression of its own particular environs (whether social or natural), every theory also has the potential to enable us to transcend (go beyond) its origins. This is why it is worth thinking at all. As will become obvious, this leads to a very different conception of modernity from Cheney's, and hopefully a more convincing account of just how place is implicated in our discourses, lives, and values.

The Ecological Critique of Modernity and Modernism

Because of the provisos set out above there is a necessary ambiguity about whether this work is itself postmodern. All things considered, accepting such a label would probably do more harm than good since it inevitably evokes misunderstandings from those with preconceptions about its meaning. This work is however *post*-modern in the sense that it seeks to support the development of cultures that might subvert and succeed (come after) our currently ecologically and socially damaging form of life. (Cheney is absolutely right to stress that radical ecology seeks fundamental changes in the way we understand our social and natural relations.) It also sets itself unambiguously against the triumphalism of modernism (the narrative) and tries to identify and critique the modernist worldview (the dominant ideology). It is perhaps best to regard it as a "countermodernism" that seeks to radically reformulate some of the theoretical insights of modern and postmodern problematics around ecological issues.

The radical nature of the ecological critique demands that those who try to interpret, give voice to, or understand radical environmentalists' values, beliefs, and practices need to be constantly aware of the degree to which all theoretical interpretations impose limits on their subject matter. These limits often turn out to be expressions of the very cultural presuppositions that a radical critique seeks to challenge. In short, we must recognize that all our theoretical expressions are, to some degree, inscribed within the practices of our own social formation, within the past traditions, present circumstances, and future hopes of the society we inhabit. If the problems igniting environmental protests lie deep in our (occidental but increasingly global) society then, the theoretical language we have available may itself prove to be tainted by a form of life that depends for its very existence upon a continuing environmental holocaust. As Alasdair MacIntyre has pointed out, there are serious implications for theories that attempt to define solutions to a social crisis within philosophical frameworks that arise from, and whether knowingly or unconsciously, may support the very society we wish to criticize.

> The ability to respond adequately to this kind of cultural need depends of course on whether those summoned possess intellectual and moral resources that transcend the immediate crisis, which enable them to say to the culture what the culture cannot say to itself. For if the crisis is so pervasive that it has invaded every aspect of our intellectual and moral lives, then what we take to be resources for the treatment of our condition may turn out themselves to be infected areas. (MacIntyre, 1981: 3–4)

I contend that radical environmentalism does have at least some of the resources that might allow it to transcend our current situation and that it has much to say to those theorists who would otherwise be content to see it marginalized. Each of the following chapters seeks to address this marginalization in its own way. For example, chapter 1 illustrates the problems in reducing the valuation of the environment to a technical problem, an arena for the machinations of environmental economists or moral philosophers. I argue that the all-too-ready acceptance of this role by institutionally appointed "experts" often signifies an unawareness of the limitations that their intimate relation to, and inclusion within, bureaucratic social structures, imposes on their theoretical pronouncements and methodologies.

In ethics—the "expert" field with which I am most concerned—one can recognize a range of responses to our environmental crisis. Some, it must be admitted, do not even recognize the possibility of an environmental ethics let alone the relevance of MacIntyre's point. They are content to continue to view the nonhuman world as of only instrumental value and to evaluate it using cost-benefit analyses or other economistic tools. Their anthropocentric and blinkered perspective refuses even to recognize the empirically obvious—that many cultures other than our own, and many people within our own society do indeed have genuine moral concerns for our nonhuman environment. We see the destruction of whales and forests as an evil that bears comparison with crimes against humanity. To treat the value of sacred groves, rare insects, or even the tree we used to climb as a child as nothing more than a potential resource quantifiable in yen or dollars, is both an appallingly reductive misunderstanding of their complex relations to ourselves and morally on a par with being willing to sell one's grandmother to the highest bidder.

Not all theorists are so dismissive of the environment. Some, perhaps the majority of those in academic circles dealing with environmental ethics, are a little more flexible. They take on board MacIntyre's point insofar as the *content* of their ethical deliberations is concerned but refuse to recognize that the *form* of these theoretical structures might also be implicated. These people (who include figures like Peter Singer, Paul Taylor, and Tom Regan) are genuinely concerned to expand the boundaries of moral considerability beyond the human horizon, yet their philosophical methodology remains almost entirely unaltered. They simply look to the natural world for novel grounds on which we might found a theory of animal rights or argue for the inclusion of a select few nonhumans into the machinations of a utilitarian calculus. Not only does this position, which I refer to as "axiological extensionism," become progressively more impractical the further one moves from the human sphere but, as I shall argue, its form and formalism still operates to reflect and reinforce our current social structures.[24]

The predominant forms of moral theory, whether deontological or utilitarian, attempt to provide a rubric that can be used to determine right and wrong by those not intimately associated with the circumstances—that is, bureaucrats, governments, law courts, and so on. Ethics thereby becomes an abstract theoretical tool for passing judgments or evaluating actions at a distance, rather than an embedded and intimate relation to relevant others. Such formal rubrics facilitate managerial and technical efficiency, for example, in evaluating the ethical "cost" of a road development or comparing the rights and wrongs of a quarrying operation. But, in doing so, they effectively disenfranchise the moral feelings of those at ground level, promulgating and supporting the myth of a neutral rationality in the hands of professed experts impartially working in the service of society as a whole.

While supposedly neutral, this conception of ethics implies a particular understanding of the relations between theory and practice—a relation by which theory claims to encapsulate and represent the essential features of moral activities and then reapply them. Morality is thereby reduced to a series of abstract formula that can supposedly be applied to circumstances irrespective of the context of the moral claims involved. However, looked at differently, such formulae act as an ideological smoke-screen, giving the (false) impression that our moral concerns for the environment have been addressed, weighed in the incorruptible balance of rational thought, and found wanting.

These explicit and formalized systems have come to colonize and dominate the modern life-world. And, I argue, in adopting this form current ethical theory operates as yet another kind of positive philosophy—that is, a philosophy that supports rather than subverts the current status quo—philosophy as an instrument of social management rather than as an expression of genuine moral concerns.[25] In Zygmunt Bauman's words, contemporary ethics "embark[s] . . . on an arduous campaign to smother the differences and above all to eliminate all 'wild'—autonomous, obstreperous and uncontrolled—sources of moral judgement" (Bauman, 1993: 12). Formal ethics misconstrues and eviscerates our actual moral feelings in order to incorporate them, as pale shadows of their former selves, into a hierarchical society where others take decisions for us. In this way axiological extensionism, no less than its more blatantly instrumental and anthropocentric counterparts, defuses environmentalism's radical critique of Western society by marginalizing those who try to speak with a "different voice."

Chapter 2 turns from philosophy to social theory in order to investigate further the links between modernity, theoretical frameworks, and ethics. In particular it focuses on the ethical theory of Emile Durkheim as both the single most important influence on social theoretical understandings of ethics and as an exemplar of how theory inevitably reconstitutes "the ethical" according to its own presuppositions. Durkheim provides a valuable account of the changing nature of ethics from premodern societies, which, he claims, exhibit a mechanical solidarity, a moral consensus based upon the similarity between individuals' social roles and values, to modern society, characterized by a much greater division of labor. This specialization in roles inevitably induces differences between individuals' values. Modern society exhibits what Durkheim terms "organic" solidarity, a form of moral regulation based upon the necessary interdependence of each specialized role-taker upon others. In organic society morality has to take on a more abstract form since its purpose is to facilitate communication between individuals whose specific values may conflict. In this way Durkheim can help explain exactly why the forms of moral theory criticized in chapter 1 have such a powerful influence over our patterns of thinking and institutions today. Talk of abstract categories like "rights" and "utility" arises precisely because these are supposedly sufficiently disengaged from individuals' disparate personal concerns to provide a common ground upon which all members of society can agree.

However, this explanation comes at a cost. Durkheim is, I claim, not sufficiently reflexive and fails to see that his theoretical intervention has reconfigured "the ethical." Durkheim's normative and functionalist characterization of ethics regards it as a kind of social glue, serving to resist change, fix current social relations in place, and ensure their reproduction from generation to generation. Such functionalism means that the ethical loses all specificity as a relatively autonomous category of social relations and is recognized only in its communicative and prescriptive manifestations. Its inspirational vitality is smothered and the phenomenology of ethical feeling ignored except insofar as it is regarded as a necessary epiphenomenon of moral regulation. The heartfelt aspects of ethics—of love, care, sympathy, and the like—are turned into a social currency to facilitate tradeoffs between people with different interests and aims. Importantly, Durkeim's normative account also neglects the possibility of ethical critiques of current social relations and ethical arguments for social change. Radical ecology relies, to a large extent, on precisely such ethical critiques.

Chapter 3 takes this analysis further examining some of the problems that arise when current sociological and political perspectives that are explicitly anthropocentric, attempt to address environmental critique. I suggest that while radical environmentalists may find much that is worth drawing upon, the major sociopolitical paradigms are unsuitable vehicles for reappraising the relationship between nature and culture. This is perhaps unsurprising since the very idea of sociology as a separate disciplinary field is dependent upon the reification of a nature/culture dichotomy. Theoretical debate has, until recently tacitly accepted this dichotomy and has confined itself to epistemological and methodological problems associated with studying society rather than challenging the category of "society" per se. Thus debates have arisen between those who advocate a unified method encompassing the social and natural sciences—which obviously includes positivism—and those, like Weber, who argued for a separate discourse of interpretative understanding to take into account the intentionality of human activities. Despite their erstwhile differences, each in their own way accepts the existence of the "social" as a necessary and objective category. They see themselves as studying the same *thing* but using different methodologies. In other words they implicitly accept disciplinary boundaries that are united not so much by what they include as topics for study but by what they agree to exclude, that is, nature.

If positivism and interpretative sociology fail to meet environmentalism's needs, Marxism, that other founding strand of sociopolitical theory, is equally unsatisfactory. This too comes as no surprise since, despite its overt materialism and some recent and imaginative textual exegesis, it is difficult to find anything in Marx's voluminous writings that gives anything but an entirely passive role to nature.[26] Nature remains the raw material for human economic activities and Marx was neither a green before his time nor was he a political Nostradamus who foresaw our current environmental predicament.

Given these historical predilections, social theory hardly seems to be fertile ground for a radical reappraisal of the nature/culture interface. However, just as chapter 1 finds much of value in Weber's description of modernity, bureaucracy, and instrumental rationality and chapter 2 borrows from Durkheim's work on ethics, chapter 3 attempts to recuperate something from Marx by appropriating and adapting his productivist epistemology and critically appraising the dialectic. The intent then, is in each case to reformulate certain insights from social theory while recognizing that theory cannot be just about society.

This somewhat eclectic approach mirrors contemporary social theory insofar as it too lacks a monolithic disciplinary consensus and comprises a number of heterogeneous strands of thought that jostle for attention and credibility. Many strands have something to add to the development of an environmental ethics. Feminist critiques of dominant theoretical paradigms are particularly relevant and later chapters draw extensively on feminist ethics, standpoint epistemologies, and difference theory. The work of feminists like Carol Gilligan, Judith Butler, and Luce Irigaray, together with that of ecofeminists like Val Plumwood, are invaluable in attempting to subvert the nature/culture divide and reformulate an environmental ethics.[27] This is partly because the situation facing environmentalism is in many ways analogous to that confronting feminist social theorists. Feminist sociologists like Dorothy Smith have argued that feminism requires a sociology *for* women rather than *of* women since the latter accepts the possibility of attaining an objective sociological stance (Smith, 1987). This is purposively contentious and asks theorists to reassess their relationship with the "objects" of their study. Similarly, I would argue that one cannot just have a sociology and philosophy *of* the environment but should have theories that are openly and passionately *for* the environment.

Chapter 4 takes the debates between radical environmentalism and social theory further, examining recent conflicts between adherents of deep ecology and social constructivism. The power and the promise of deep ecology is seen, by its supporters and detractors alike, to lie in its claims to speak on behalf of a natural world threatened by human excesses. Yet to speak of nature as something worthy of respect in itself has appeared increasingly difficult in the light of social constructivist accounts of cultural and historical variations in our concepts of nature. Constructivism contends that, in McNaghton and Urry's words "there is no singular 'nature' as such, only a diversity of contested natures; and that each such nature is constituted through a variety of socio-cultural processes from which such natures cannot be plausibly separated" (McNaghton & Urry, 1998: 1). While there are various forms of social constructivism, each of which places emphasis on the significance one or more aspects of social practices (discourse, culture, etc.) in producing our "lived reality," they all seemingly threaten to undermine the biocentric values proclaimed by radical environmentalists like deep ecologists. If values are no more than human creations and vary from culture to culture then talk of nature's *intrinsic* value seems unsupportable. Deep ecology has so far been loath to take constructivism's insights seriously, often retreating into forms of biological objectivism and reductionism to support their arguments. Yet, I argue, as a form of radical environmentalism deep ecology actually has much in common with, and to gain from, certain kinds of constructivism and can add a new dimension to constructivsm's own critique of current ideologies.

Chapter 5 returns to the themes of the earlier chapters and tries to account for the nature of environmentalism's aversion to formalized and codified rubrics by examining the values expressed in radical environmental protests—in this case the antiroad movements in Britain. These values are characterized as a form of antinomianism, the key constituents of which might include the following. First, the *rejection of political authority,* in particular as it is embodied in the law and its associated institutions. Second, the *rejection of all moral authority* and a consequent refusal to accept the imposition of moral laws whether secular or religious. Third, a deep seated suspicion or even outright *rejection of rational authority* insofar as it claims to be the sole arbiter and administrator of our lives. Fourth, an emphasis on the free individual as being responsible for creating their own space of engagement with the world and last, but not least, a utopian belief in a New Jerusalem, a new world without an order and without hierarchies. In short, environmental antinomianism sets liberty and love against the law, and inspiration against formal reason thereby expressing a fundamental critique of contemporary society, contemporary morality, and current theory.

Recognizing this antinomian and an-archistic side to environmentalism is important since it helps to disentangle it from political and moral misunderstandings and misrepresentations. Politically there is an attempt to reduce radical environmentalism to a form of so-called ecologism—usually defined as replacing social laws with a willing submission to "natural" laws. Morally there is an attempt to translate the ethics of such actions into already existing frameworks of rights, justice and utility. Both of these "rationalizations" want to create axiologies, to *define, to place boundaries on, and express the essence of* activities that, I argue, explicitly set out to subvert all such boundaries and reifications.

This argument is further developed in chapter 6, which suggests a possible alternative for theorizing a radical environmental ethics. Again inspired by recent ecological protests, it attempts to articulate a radical environmental ethos voiced in terms of a spatial metaphorics, "an ethics of place." It reconstitutes ethics spatiotemporally in order to counter the current enclosure of the moral field within economistic and legal-bureaucratic frameworks and institutions. This alternative ethical paradigm explicitly recognizes the importance of locality and context and, at the same time provides a language more suited to expressing the values of those forms of life associated with radical environmentalism. This ethics of place, might, I contend, be better able to comprehend the value in being close to nature.[28]

Of course, the creation of a new ethics is, in the last instance, a social rather than a purely philosophical project. My theoretical appropriation of environmental protests only provides the briefest outline of a potential harbored within the emergent culture of radical ecology. There is also the ever-present danger that my own theoretical approach may misconstrue or place new restrictions upon radical environmentalism. All theoretical discourses, "problematics," frame and highlight certain questions, interests and presuppositions, excluding certain possibilities from consideration, making some concepts central and others peripheral. In doing so they create a position, "a particular unity of a theoretical formation," which "binds" those who use it.[29] A problematic is not just a theoretical tool to be applied to the world but the framework within which problems develop and proposed solutions are judged. A problematic both creates and emphasizes particular theoretical relations to the world. The boundary between my appropriation (using for special purposes) of ecological protests and their expropriation (taking away) has to be constantly born in mind. Yet, despite this danger, I hope that this ethics of place is capable of expressing something of the multidimensionality of radical environmentalism's unashamedly utopian project.

If radical environmentalism does indeed have some affinity with antinomianism and even anarchism then this might be thought to pose problems for both ethical theory and practice. After all, anarchism is associated in many peoples minds with a radical individualism that regards each of us as a law unto ourselves; a view of liberty that easily verges on moral libertinism and a nihilistic conception of politics. These traits seem to be the very opposite of what usually passes for the subject matter of ethical activity and theory. How can individual autonomy be squared with concern for others, and nihilism equated with those innovative and imaginative expressions of hopes for the future that characterize contemporary environmental activism? Chapter 7 tries to address some of these issues by offering a reappraisal of writings on the ecological self in the light of the work of Judith Butler and Luce Irigaray. Irigaray provides an account of an ethics of (sexual) difference that might also illuminate our ethical relations to natural "others." In many respects this is the key to the whole of this project. The hope here is to link together concepts of the ecological self and an environmental ethos, an ethics of place, in order to demonstrate how radical environmentalism might re-envisage morality as an expression of a heartfelt but uncodified *modus vivendi*. This term is used by Arne Naess (1979) and refers to the development of what chapter 8 characterizes as a *sensus communis,* a practical moral sense of the needs and value of our fellows and surroundings that is constitutive of genuine self-identity. A *modus vivendi* is a mode of being that gives due, but not formally equal, respect to all those things recognized as parts of a community, despite and often because of their differences to ourselves. It requires a learning to *be* and *let be* in natural places. I suggest that this might explain why, rather than looking for abstract formulae to encapsulate the morality of environmentalism, we should understand it as the internalization of an environmental ethos and the externalization of an embodied ecological *habitus*.[30] In this way I hope to come full circle and show how taking nature seriously can and must have profound effects not just on the way we think (our theories) but on the way we live (our practices).

1

Against the Rationalization of Environmental Values

I want to begin by making some general remarks about the approach that I will take to ethics without offering too much in the way of justification for them at present. I hope that, as chapter follows chapter, the rationale behind my adoption of a sociotheoretical rather than a purely philosophical perspective will prove increasingly compelling. Thus, although this chapter initially focuses on examples of those philosophical arguments that have tried to take various features of the natural environment into account, it does so largely in order to bring out certain shared but often unacknowledged background assumptions characteristic of their social origins in modern occidental culture.

Given this sociological approach, it would be difficult if not impossible to start by presenting a definition of ethics per se because the moral values of any given society or social group differ markedly, intertwined as they are in complex ways with that culture's prevailing forms of life.[1] How we live obviously affects what we value and how we come to value it. Clearly, values also work reciprocally to inform how we live. Anthropology shows us that there can be no pregiven limitations on what subjects ethics might be called upon to address or on how it comes address them. A recent collection includes essays on, among other things, "the absolutism of landownership in an English village," "the moralities of Argentinean football," and the wonderfully entitled "'I lied, I farted, I stole . . .': Dignity and morality in African discourses on personhood" (Howell, 1977). Such examples prove that questions of rights and wrongs, of sin and denial, of authority, love, happiness, health, and even farting, are the subjects of culturally and historically variable moral discourses. Yet, despite such evidence, the extent of ethics' entanglement with society often goes unrecognized by many of those who seek to provide a theoretical account of its activities.

I will argue that the frameworks we use to think about ethics are no less culturally variable than the subject matter they are called upon to address. Both the *content* and *form* of ethical discourses are products of particular times and places. For example, Richard Tuck has shown how the discourse of 'natural rights', which today holds global institutional sway, is a recent development with its origins in fourteenth- to seventeenth-century Europe (Tuck, 1977). In his three-volume "history of sexuality" Michel Foucault goes further still, charting the interconnections between sexual mores and morals, changing conceptions of the self and differing philosophical discourses and practices in ancient Greece (Foucault, 1985; 1986; 1990). In this way he exhibits some of the myriad connections between ethical and social relations. For these reasons, the *articulation* of an ethical discourse might best be understood, as the word suggests, as a way of simultaneously *expressing* and *conjoining* certain attributes of a form of life. It re-emphasizes and reinscribes the prevailing social relations, giving voice to some aspects of the social environment whilst repressing others. Every ethics helps to structure and *give shape* to what we might term a "moral field." (See chapter 8.)

Since what we count as a matter of moral concern, how we conceptualize it and how we relate (to) it varies from place to place it also seems reasonable to suppose that the more disparate the forms of life concerned, that is, the greater the social or historical differences between groups, the more difficulty there will be in correlating and mapping one "moral field" onto another. In other words these simple sociological speculations lead us to think that, to avoid an error of conflation, we may have to speak of moralities (plural) rather than morality (singular). Yet moral philosophy has often presented ethics as some thing that can be universally defined and delineated, as though that which comprised the specifically moral element of any society, activity or thinking was readily distinguished and agreed.[2] This assimilation of differences between moralities is, as Alisdair MacIntyre (1966) has remarked, a "fatal mistake." Moral systems exhibit immense variety. They share no essential features but are recognized only by their, often distant, family resemblances to each other. To obscure these vital differences by tokenistic fiat is an error symptomatic of attempts to analyze moral concepts in isolation from their actual contexts.

If these initial sociological points are accepted, then ethical philosophy, despite appearances, should not be regarded as an attempt to capture in discourse that which comprises the essence of moral behavior per se, for no such essence exists. Rather, every ethical discourse expresses and constructs a different moral field, draws different boundaries around morality, writes or speaks of a different cultural context. Of course, overlaps do exist between forms of life that allow different ethical systems to communicate with each other up to a point. But we should remember that ethical debates are not simply about the kind of world we should inhabit but the kind of social environment we actually do inhabit. They are signs of genuine disagreement about the boundaries of morality itself, meetings of conflicting worldviews and social structures, of different ethical *spaces* and *terrains*.

Ethical theory should not then, on this reading, be regarded as a quest for transcendental moral truths and principles, but nor can it be reduced to a descriptive science, a search for accurate anthropological representations of differing cultures' feelings and values. It is neither a conceptual quest nor a branch of social science but something both more active and mundane. Every ethics is a *shaping* and *enframing* of the social and natural environment, it sculpts and composes a different world within the material constraints and possibilities afforded by its own particular time and place.

By speaking of moral fields and of ethical spaces, I hope, over the course of this work, to point out the limits of our dominant moral discourses; to show what they do and do not encompass. The ethical discourses that dominate our current agendas, whether they be utilitarian or rights based, are themselves the relatively recent products of a time and place we might refer to as "modernity." They reproduce the particular social and natural relations pertaining to industrial capitalism.[3] They have helped shape today's world and their currency, prestige, and almost unchallenged hegemony in philosophical discourse is a result of their fitting with those contemporary mores that they are themselves, in part at least, responsible for producing. Modernist discourses like utilitarianism may have opened up previously unrecognized moral dimensions but they have also, of necessity, distorted and even annihilated those previously inhabitable moral spaces they played a part in superseding.

Radical ecology presents a serious challenge to those aspects of modernity that it regards as complicit in the increasing degradation of our natural environment. It rejects many elements of our contemporary "forms of life." It therefore follows that environmentalism must find other ways to articulate its ethics because the established forms of ethics, insofar as they are representations and embodiments of modernity, will inevitably distort or exclude the values of critics who live or envisage a different form of life, an alternative ethos. This initial chapter aims to indicate the manner in which the dominant ethical discourses of modernity have enclosed the moral field. This enclosure is characterized as an extension of that process of rationalization that sociologists from Weber to the Frankfurt School have regarded as central to modernity's project. Later chapters will sketch, if only tentatively, an alternative ethical paradigm, one that explicitly recognizes the importance of natural locality and context and, at the same time, might provide a language more suited to expressing the values of those forms of life associated with radical ecology. The ultimate aim of this work as a whole is to *re-place* contemporary ethical discourses in both senses, that is, to *put them into* the context of their specifically "modernist" origins and to offer an example of an alternative ethical discourse which might *supersede* them.

As the introduction indicated, and the remarks above have emphasized, the creation of a new ethics must always be founded in a social rather than a purely theoretical project. The social project that is radical ecology has, at present, precious few theoretical allies. But the following chapters represent an unashamed attempt to develop one possible theoretical alternative of a kind that might be termed, in Ernst Bloch's (1995) sense, a "philosophy of hope," based in the critiques of modernity and its principles voiced by radical environmentalists and given shape in their activities.

However, before I can turn to a critique of current attempts to give a moral voice to aspects of our natural environment, I need to address those who do not even deem nonhuman nature worthy of our *ethical* concern, remaining wedded to an entirely economic and instrumental valuation of the environment. Here the environment is simply regarded as a "resource," albeit one that might, given the right economic climate, be used in a "sustainable" manner.

Environmental Economics

Neoclassical economics might be regarded as that instrumental form of rationality that most actively opposes the ethical evaluation of the environment (at least insofar as ethical concerns are seen to lie outside of the gratification of personal preferences). Excluding a few honorable exceptions, the growing discipline of environmental economics is engaged in nothing less than a blatant attempt to reduce the complexities of human/environmental relations to a uniform economic metric.[4] The current tendency toward an overt economization of values is exemplified in such documents as David Pearce and colleagues' (1989) *Blueprint for a Green Economy,* originally a report for the British Department of the Environment, and Wilfred Beckerman's (1990) *Pricing for Pollution.* According to these economists money is *the* institutional form best able to express human interests.

Pearce's *Blueprint,* and its numerous successors, argue for the integration of environmental concerns into economic policy. The authors regard economics as a theoretical methodology able to express the values underlying competing preferences, for example preferences for more roads or an unspoilt landscape. It is claimed that all "rational" people would wish to maximize their satisfaction. However, in a finite world there are limits to preference satisfaction. To maximize social utility we must therefore be able to measure people's preferences; money is presumed to provide this measure.

According to these environmental economists our past environmental problems have not been caused by economic rationality but rather by distorted accounting procedures and by the lack of a marketplace for environmental services and concerns. Once this structural problem has been rectified, industrialists and developers will change their ways and we may leave environmental decisions largely to the free market. Accounting methods that previously regarded environmental damage as an "externality" simply need to be amended so that the value of habitats, species, and so on can be made economically tangible. The environment's true economic value is to be determined by methods of "contingent valuation." People will be presented with a hypothetical market and asked how much they are "willing to pay" (W.T.P.) to protect any given environment or alternatively "willing to accept" (W.T.A.) in recompense for the loss of environmental quality. These figures can then be entered into a cost benefit analysis alongside other more standard economic data. Thus, the economists say we have a "rational" way of making decisions on environmental issues that takes all important factors into account.

However things are not so simple. First, as many critics have pointed out, the figures that result from W.T.P. and W.T.A. surveys vary widely. People generally ask for much higher compensatory sums for losses than they are willing to pay for similar gains. Can both methods be appropriate if their results differ so much? Secondly, such monetarization seems likely to reflect the depth of one's pocket rather than the depth of one's feeling. As one environmental economist critical of the overextension of contingent valuation, Donald McAllister (1980), has remarked, "cost and benefit are typically added without attempting to adjust for the likelihood that a dollar is valued differently by people at different income levels."

More importantly, Pearce and his fellow economists have largely ignored the fact that, as Mark Sagoff (1989) points out, such studies in contingent valuation of environmental matters typically meet with a high proportion of protest bids and outright rejections of the hypothetical scenario.[5] Sometimes up to 50 percent of those surveyed refused to take part in the survey or required huge sums of money as compensation. These protest bids are simply ignored by the analysts, who seem blithely unaware that many people do not share their conviction that environmental concerns can be expressed so easily in financial terms.[6] As McAllister (1980: 143) puts it, "[f]or years certain proponents of CBA [cost benefit analysis] have been selling it as a completely comprehensive evaluative method, capable of incorporating in its grand index all the factors important to public decisions. . . . But some of its serious limitations are inherent in its fallacious premise that all important human values can be adequately represented by money."

The valuation techniques utilized by environmental economists make a number of questionable assumptions about what constitutes a "rational person," about the nature of our values, and about our relations to our social and environmental surroundings. In the economists' eyes we become nothing more than calculating bundles of self-interested preferences; economic morons completely lacking in ethical or aesthetic sensibilities. But there are, as Mark Sagoff remarks, many areas of human life where we recognize that consumer preferences do not have a place. Few people would suggest that the outcome of murder trials should be decided upon a criterion of willingness to pay, and only the craziest of economists would argue for and against such issues as abortion and slavery on economic grounds.[7]

Pearce and Beckerman's arguments explicitly assume that economic rationality is a neutral instrument for reaching policy decisions. This is compounded by a view of all values as reducible to a one-dimensional and quantifiable metric. To deny the efficacy and universality of economic evaluation and to insist on the import of ethical considerations, is for some economists enough to warrant being labeled irrational. For example, Beckerman contrasts his own "cool" and "logical" views with the "emotionally-charged reactions of the anti-growth school," (Beckerman, 1990: 29), that is, environmentalism. To even mention ethical considerations, or question economic assumptions is, he thinks, "an emotional over-reaction to some of the obvious disamenities [sic] of modern life" (Beckerman, 1990: 22). Beckerman holds that those who oppose such accounting methods "are no doubt motivated by other considerations. [And he claims] in the absence of any opportunity to subject them all to psychoanalysis it is not possible . . . to speculate on their inner motivations" (Beckerman, 1990: 29). The implication being that those of us who object to being characterized as selfish bundles of personal preferences must be mad.[8]

Neither Pearce nor Beckerman think it necessary to question the use of nature as a human *resource*. But this is precisely the point of contention with many environmental ethicists who would all question this across-the-board application of economics. Those of us who feel concern for the environment regard our feelings as akin to those that we might have for other humans. It simply is not possible to work out an economic value for someone one loves; to treat her or him as a resource rather than a person. If asked the "value" of our grandmother we wouldn't institute a hypothetical market for aged relatives. We would quite rightly see any attempt to confine the question of value to monetary terms as, at the very least, inappropriate if not downright evil. Nor are the values many of us place on the existence of rainforests and whales reducible to dollars and yen. Environmentalism requires the widening of our ethical and aesthetic concerns, not just our preferences as consumers.[9] The questions involved are not just about the allocation of resources but about morality and politics, about the very notion of treating nature as a resource.[10]

There is another important, but seldom mentioned point here. The principle of conservation, the action of keeping from harm, decay, loss or waste, that is, of maintaining spaces where things are let be, might be interpreted ethically as the very opposite of the principle of economization. Economization, especially in neoclassical frameworks, insists that values are entirely dependent upon the recognition of things exchangeability, of their inclusion or potential inclusion within systems of exchange, that is, markets. The idea that some things might not be exchangeable, that they are not for sale at any price (even potentially) simply and literally cannot be *accounted* for by economics.[11] Contingent valuation depends upon the extension of the market into all spheres of life, of making things quantifiable and therefore tangible, a tangibility that depends crucially on the recognition that their transformation or exchange is at least possible. On the other hand ethics and conservation both frequently have to insist on the ultimate intangibility of certain values, on the recognition that some things are unique, irreplaceable, nontransferable, and so on, but nonetheless held dearly. (See chapter 7.) This difference perhaps accounts for the almost schizophrenic attitude exhibited by many conservation bodies who accept the current commodification of nature at the risk of it quite literally costing the earth.

In this sense then it seems obvious that to accede to the economization of nature is to merely extend the commodification of the life-world. Far form being value-free this economic approach replaces the complexities of political, ethical, and aesthetic values with a simple-minded economism. But, despite their occasional use of resource-oriented rhetoric, most radical ecologists regard the conservation of our natural environment as a matter for moral concern and not just a matter of human utility. Rainforests might often be referred to as gene-banks, potential resources for sustainable development, oxygen factories, and so on. But, though in some sense they may be all of these things, to justify their preservation by reference to these roles is to accept the language and rationale of their exploitation. These are expressions of anthropocentric attitudes toward nature and concrete examples of the imposition of managerial and financial constraints upon nature. Just as in our present bureaucratic/consumer society all has to be managerially approved and financially profitable, nature too, it is often argued, needs to justify its continued existence on the same grounds. Though the defense of wild places by such means may be successful as a short-term expedient, to justify their preservation only, or even primarily, in these terms is tacitly to accept the status quo and the ultimate hegemony of human self-interest.

The long-term consequences of such a policy are likely to be disastrous. If the fundamental reasons given for preserving habitats are those of human utility, then, whenever and wherever the balance of utility, measured economically, favors habitat destruction this will occur. Once destroyed they can rarely be replaced. Bit by bit the wilderness is eroded until all that remains are a few curios, remnants of what once was, to be stared at and picked over. This is not just idle speculation, it accurately describes our current situation.

Axiological Extensionism: The Expanding Circle

If the economization of values represents an ever-present danger to the environment can ethical theory do any better? There has been no shortage of attempts over the last twenty or so years to develop moral frameworks for thinking about nature. However, any moral problematic developed to deal with human interests obviously requires some degree of adaptation in order to be applied to nonhuman realms. For example, a straightforward moral utilitarianism would fare no better at expressing environmental values than homologous systems of economic utility. The moral utilitarian may have a different metric of utility, a hedonistic calculus instead of a monetary calculus, but their methodology would be subject to identical drawbacks insofar as giving a rationale for the preservation of wilderness is concerned.[12] Whenever the balance of utility, measured as maximized human happiness, favored habitat destruction or species extinction it would be deemed morally right.[13] Much debate in environmental ethics has therefore centered on the need to expand the category of moral considerability (e.g., Stone, 1974; Goodpaster, 1978; Brennan, 1984).

One such attempt to expand morality claims that if we look at the history of Western society there seems to have been a unidirectional expansion of the bounds of "moral considerability," from the immediate social group to ever widening categories of moral objects. Peter Singer has noted that a popular metaphor for describing this broadening of ethical horizons is that of the expanding circle. A typical example he quotes comes from Lecky's "History of European Morals" first published in 1869. "[B]enevolent affections embrace merely the family, soon the circle expanding includes first a class, then a nation, then a coalition of nations, then all humanity and finally its influence is felt in the dealings of man with the animal world" (Lecky in Singer, 1981).

This metaphor of the expanding circle encapsulates a specifically modern idea of a linear historical progression, a continual moral improvement as humanity becomes increasingly "civilized." Singer's own book *The Expanding Circle* both elucidates and epitomizes this approach. His thesis is that ethics originated in forms of biological behavior such as kin selection and reciprocal altruism, whereby apparently altruistic acts of individuals are explained by their role in increasing the genetic contribution of that individual's genome to the gene pool of the next generation. A mother shares 50 percent of her genes on average with an offspring. Put in its crudest form, those mothers who die saving more than two offspring will be selected for. Thus altruism as a feature seems amenable to explanation in terms of so-called selfish genes (Dawkins, 1989).

The altruistic faculty, according to Singer, comes to take on a new form for humans because of our endowment with language and rationality. Justifications of actions affecting the wider community come to be given in terms of reasons. For example, I may justify my claim to a greater than average share of the food on the basis that I do more work than most. This might be accepted but then someone points out that some other person does more work than I do, and so is entitled on this basis to more food still. Once utilized, this form of rational argument suggests that we "cannot get away with different ethical judgements in apparently identical situations" (Singer, 1981: 93). Certain "rational" considerations can call into question previously held prejudices about the limits of moral considerability. For example, if it is right to help person A in a given situation then why not person B?

Altruistic tendencies had in the first instance, we are told, only extended to the immediate family, or our own group, but the "autonomy of reasoning" (Singer, 1981: 113) entails a logic whereby the boundaries of our own expand to the next largest community with which we identify.[14] Perhaps this community is a social class or a race but once such an extension of moral considerability has been justified then its boundary too is in turn open to questioning. Why for example should skin color be regarded as a criterion that could justify excluding someone from moral consideration? Viewed in this way, the history of morals comes to be seen as an increasingly enlightened view about those we conceive of as having affinities to ourselves. Like the layers of an onion the boundaries of moral considerability come to overlie each other as the rational justification for each is formulated and then challenged.

Eventually we reach a stage where claims to moral status are justified in terms of features of something called "human nature"; perhaps the possession of a rational faculty itself. This stage is equivalent to the roughly Kantian position, that if I am morally considerable because of my rationality then all rational beings must be so considered. This being so, individual members of different races, sexes and so forth apparently obtain equal moral status (unless of course we can find reasons for doubting that all sections of humanity are equally rational).[15]

But why should the policy of extension stop at the level of species? In a famous quotation, Bentham points out both the drawbacks in relying on rationality or language to delimit moral considerability. He suggests instead that ability to feel pain or pleasure is the appropriate moral arbiter: "It may one day be recognized, that the number of legs, the villosity of the skin, or the termination of the *os sacrum,* are reasons equally insufficient [to skin color] for abandoning a sensitive being. . . . What else is it that should trace the insuperable line? Is it the faculty of reason, or perhaps, the faculty of discourse? But a full-grown horse or dog is beyond comparison a more rational, as well as a more conversable animal, than is an infant of a day, or a week or even a month old . . . the question is not, can they reason? nor can they talk? but, can they suffer?" (Bentham, 1907: 311).

This is indeed the position that Singer takes, claiming that this is the outer layer of the onion. The difference between sentience and nonsentience is not, says Singer, a morally arbitrary boundary in the way that species differences are. Unfortunately, so far as environmentalists are concerned, Singer's position has an obvious drawback. Drawing the line at sentience excludes most of the animal kingdom and certainly plants, waterfalls, and whole ecosystems from moral considerability. However, ingenuity can fill out this notion of rationally argued affinities in still other ways. Instead of using shared natural characteristics—genes, skin color, sex, human nature, or even sentience to dictate moral boundaries one could refer to shared interests.

Now, if interests are of critical importance, the outer layer of moral considerability will be bounded by an ability to possess interests. Singer believes that the capacity to possess interests is co-extensive with sentience, but others have a wider perspective. Why should plants not have interests in obtaining enough water and nutrients? Thus, for philosophers like Robin Attfield plants too find a place within the expanded circle. For him the interests of nonsentient beings lie in "their flourishing or their capacity for flourishing after the manner of their kind" (Attfield, 1991: 154).

Paul Taylor (1986) similarly proposes a "biocentric" theory of environmental ethics where the outer layer of moral considerability is to be determined by a thing's ability to possess a "good-of-its-own." To have a good-of-its-own the object must be capable of being harmed or benefited as a teleological center of life, having its own species-specific goals. The goals of an organism are realized when it has successfully maintained "the normal biological functions of its species," thus developing to its full potential. A butterfly species, for example, has a life cycle from egg to caterpillar to chrysalis to imago. To stop any individual butterfly from playing each of these roles would constitute a harm to it. Having a good-of-one's-own is then a necessary, but not sufficient condition of moral considerability (of having inherent worth in Taylor's terminology). This distinction between things that have and that lack a good of their own equates, according to Taylor, to that between the living and nonliving and constitutes the principle that marks the outer boundary of moral standing.

To summarize the argument so far: These approaches, which we might term axiological extensionism, that is, the expansion of formal systems of allocating values, present us with a succession of features or capacities that are supposed to determine the bounds of moral considerability. All previous boundaries as they become superseded are seen to have been mistaken, their core justifying principle being too limited in scope. They were based on the wrong objective essential characteristic: that characteristic, which has the role of carrying, or at least grounding, value. But now the ethicist faces a serious dilemma. For, as the boundary principles become less and less specific to take account of the wider categories of ethical objects we wish to countenance, this form of rational argument brings with it a new and more expansive egalitarianism.

If, for example, possession of interests is the criterion used there seems to be no overarching reason why the interests of one type of organism should have more importance than any other. All things capable of having their interests benefited or harmed are equally considerable whether aphids, dandelions, or humans. This extreme position would be held by very few. But, on the other hand, to relate everything to similarity of interests with humans seems unjustifiably prejudiced. Faced with the possibility of widespread natural egalitarianism most axiological extentionists backtrack and busy themselves constructing rational justifications for their prejudice in much the same way as others had previously tried to exclude various sections of humanity from equal consideration. For example, Robin Attfield, (1991) having extended the boundary of moral considerability to those things capable of possessing interests, then constructs a "rational" justification that, in effect, severely limits the degree of consideration we can actually give things to their degree of similarity to humans.

Similarly, Taylor is explicitly concerned to promote natural egalitarianism as the heart of his biocentric perspective. He states, "[a]ll animals however dissimilar to humans they may be are beings that have a good-of-their-own" and "all plants are likewise beings that have a good-of-their-own" (Taylor, 1986: 66). He claims that the first thing we do when we accept the biocentric outlook is to take the fact of our being members of a biological species to be a fundamental feature of our existence. We do not deny the differences between ourselves and other species, any more than we deny the differences among other species themselves. Rather, we put aside these differences and focus our attention upon our nature as biological creatures; "we keep in the forefront of our consciousness the characteristics we share with all forms of life on Earth. Not only is our common origin in one evolutionary process fully acknowledged, but also the common environmental circumstances that surround us all. We view ourselves as one with them, not as set apart from them. We are then ready to affirm our fellowship with them as equal members of the whole Community of Life on Earth" (Taylor, 1986: 101).

Such is his theoretical standpoint, but when it comes to the practical implications of this policy for human interaction with the environment he is less candid. All that this egalitarianism practically requires is that; "*certain* habitats used by wild–species populations are not destroyed, and *some* wildlife is given a chance to survive alongside the works of human culture." "Animals [says Taylor] are not of *greater* worth [than humans] so there is no obligation to further their interests at the cost of basic interests to humans."[16]

But surely there are cases where if *equal* moral status is to count for anything, the basic interests of animals and plants will outweigh those of humans. Indeed since Taylor's theory gives inherent worth to microscopic individuals almost every act we do becomes of immense moral import, harming and destroying millions of our fellow citizens. In spraying a crop we destroy vast quantities of insects, fungi, and so on, all supposedly on an equal footing with ourselves. Taylor chooses to ignore the potentially restrictive nature of his thesis and instead makes some extremely bland generalizations about living in harmony with nature.

In effect Taylor fails to distance himself from the anthropocentric attitudes he originally claimed to be resisting.[17] Despite the erstwhile philosophical differences between Taylor and Singer their philosophical methodologies have much in common. Both are engaged in the development of ahistorical, acontextual and essentialist moral axiologies, theories of value that regard values as having objective standing because of the nature of the objects they "attach" to. Both posit that there are certain essential features of humans, and some animals that serve to ground values. Both also unquestionably accept the "autonomy of reasoning," regarding reason as a neutral arbiter between all values or as an impartial tool capable of ascertaining the scope of application of all values. As I hope to show, neither recognizes the influence of prevailing social circumstances on the form and content of their own arguments and presuppositions.

Extensionism and Environmental Holism

Before pursuing these points further I want to look at a closely related way of expanding ethical consideration.[18] The philosophical rationale for extending moral considerability has so far depended upon the sharing of certain features or capacities judged to be of moral import. A different method might depend not on sharing anything with nature but upon sharing in nature itself. J. Baird Callicott who takes a holistic stance based on his interpretation of quantum physics provides one example of this form of argument. According to Callicott our apparent individuality and isolation from nature is mistaken, for, at the level of quanta we are actually continuous with the world: Callicott endorses Alan Watt's Buddhist sentiment that "the world is your body" (Callicot, 1985: 274). If this is the case, then Callicott thinks we can dismiss arguments about the intrinsic value of different attributes, we need only posit that the self is valuable: "nature is intrinsically valuable to the extent that the self is intrinsically valuable." Environmental degradation is thus to be seen as an attack on my extended person: "the injury *to me* of environmental destruction is primarily and directly to my extended self, to the larger body and soul with which I am continuous" (Callicott, 1985, 275).

This form of argument is immediately open to a number of criticisms because humans do not operate ethically at the level of quanta. The fact that we are one with nature at this level gives us no ethical guidance at all, for so too are murderers, logging companies, and industrialists. This is not to say that the perception that we form a part of a greater whole will always be morally insignificant. Such a view may for example lead to the valuing of nature as a whole system. The acknowledgment of holism may be central to particular ethical ideals in other ways, as it was for the stoics and Spinoza.[19] However, by itself, Callicott's holism cannot give us any ethical guidance. To live in the world we need to act differentially to parts of it. We have to relate on a human scale with whales, mountains, and other humans. Ethics is about the resolution of conflicts between ourselves and about our relationship with our environment. The existence of physical links does not necessitate that conflict will cease.

Further, in terms of quantum physics it is very difficult to talk about environmental destruction at all. The destruction entailed when a beefburger is produced via a circuitous route from forest, to grassland, to cattle, cannot be expressed in terms of quanta. It can only be expressed in terms of the forest itself. Fundamental ethical dilemmas are left entirely unaltered by this egocentric holism that could just as easily support a complete disrespect for the surrounding world on the grounds that as the world is a part of *my* body it is mine to do with as I wish.

Callicott's holism is only partly based on quantum physics. It also rests upon what might be termed ecoholism. Here the science appealed to is ecology rather than physics. Ecology, it is claimed, reveals our place in nature as a locus in an interdependent network of organisms. This interdependency should lead us to re–evaluate the worth of other natural things and see ourselves as just one among many. Again, this may be true and have important metaphorical and practical implications, but it seems far from clear that it has any *necessary* ethical implications. One could clearly grasp an ecological understanding of our place in nature and yet still treat other organisms as mere means to human ends.[20]

The problems with drawing specific moral directives from this kind of holistic extensionism are exemplified by Freya Mathews (1993) book *The Ecological Self*. Like Callicott, Mathews enlists modern physics and ecology to support her ethical holism. She begins her ethical investigations from a critical appraisal of what she terms the mechano-atomism that underlies both Newtonian mechanics and our modern understanding of the self. "The fundamental feature of mechano-atomism, logically and metaphysically speaking, then, is that it is an ontology of discrete material substances—atoms" (Mathews, 1993: 30). This "ontology cannot fail to find an expression at the social level" (Mathews, 1993: 39) a fact attested by the manner in which modernity regards the individual as an isolable, hyperseparated entity. Against these Newtonian presuppositions Mathews sets a relational conception of the subject that recognizes her interdependence on social and environmental circumstances. The existence of this so-called ecological self is supported by bringing to bear a strange (and sometimes incongruous) mixture of Spinozistic philosophy and contemporary cybernetics supplemented by "geometrodynamics" (a post-Newtonian paradigm in physics) and a contentious reading of ecological theory. Mathews argues that all these approaches implicitly or explicitly support her particular version of substance monism. "[W]e are identical with the universe: it is into its substance that the pattern that is our signature is written" (Mathews, 1993: 91).

This new cosmology suggests that we are ripples propagating in the depths of "an all pervasive medium, a medium analogous to a fluid"

(Mathews, 1993: 91).[21] But this of course raises the question of our individuality, since our distinctiveness might initially seem to be dissolved in Mathew's holistic universal ocean. She attempts to resolve this problem by recourse to systems theory, arguing that the individual is an "open system exhibiting self-regulation, homeostasis, equifinality and goal-directedness" (Mathews, 1993: 97).[22] The post-Newtonian individual can no longer be regarded as a discrete and physically bounded entity but is a temporarily persisting eddy in the cosmic stream, a cybernetic system with an ability to remain relatively cohesive in the face of changing "external" conditions.

In other words, Mathews proposes a new cybernetic criterion of individuality where an individual is characterized by an ability to maintain a dynamic equilibrium with its environment. Negative feedback loops ensure that the individual's integrity is not compromised, for example, we facilitate our continued survival in hot or cold climates homeostatically maintaining a constant body temperature through either sweating or shivering. Mathews thereby seeks to redefine the individual as a "'machine' for self-realization" (Mathews, 1993: 98). But, unlike other machines, which have their telos imposed upon them from "outside," some systems, for example, living organisms embody their purpose in themselves. Mathews goes on to argue that as soon as we recognize that each proper individual has a telos, namely its interest in its own self-perpetuation, then we must regard them as belonging to a special category of things, that is, *beings-in-themselves* all of whom "possess self-interest" (Mathews, 1993: 101). Such "self-affirming activity marks off the self-realizing being as an *agent*" (Mathews, 1993: 104) and, following Spinoza, she refers to this power of self-realization as the *conatus*.

Individuals are therefore clearly distinguishable from rocks or grains of sand since, Mathews claims, such things are not self-regulating and have no *telos* or agency. The difference between a frog and a rock is that the frog embodies, as only a homeostatic entity can, what she believes to be the essential and most important organizing feature of the universe—the possession of a *conatus*. This is not to say that Mathews limits individuality to animate objects, she does not. Her cybernetic approach means that she can extend her definition of systemic individuality to more than living organisms since certain sorts of natural systems, like ecosystems, might also be counted, in cybernetic terms, as "fuzzy individuals" with their own self-interests. This might be true even of the universe as a whole, since this too might be regarded a giant self-realizing system with its own *conatus*.

Just to complete the picture Mathews claims that this interest in self-perpetuation marks a "primitive form of [intrinsic] value" (Mathews,

1993: 104), that is, a value that is self-given and not dependent upon any extrinsic valuer. Hence, insofar as they have a *conatus*, all living beings and at least some natural systems have "intrinsic" rather than instrumental value. What is more, "[t]he *prima facie* degree of intrinsic value possessed by a particular self will be commensurate with its degree of power of self-realization, which will generally be a function of its degree of organizational complexity" (Mathews, 1993: 143). This, then, is why Mathew's ethical theory is supposed to mark a radical departure from previous anthropocentric theories of value that restricted the possession of intrinsic value to human individuals. Humanity is deposed from its pedestal as the sole exemplar of intrinsic ethical value and as the arbiter of values in the world at large since these are now inscribed in the cybernetic order of the universe itself.

But how radical a departure is Mathews' approach? To what extent does she actually overcome the limits placed on our (ethical) responses to others by the logic of anthropocentrism? It is interesting to note that, despite her avowed monism, Mathews actually introduces and defends a new dualism that establishes a "different order of being" (Mathews, 1993: 104) that "does not extend to objects such as rocks" (Mathews, 1993: 111). To this extent at least her cybernetically driven definition of individuality is actually more restrictive than the Spinozistic philosophical system she adapted since, as Mathews admits, Spinoza was quite happy to think that even rocks might have their own *conatus*. In this sense then Mathews places strict limits on who or what can count as a morally significant other.

It is also clear that by linking intrinsic value to organizational complexity Mathews reintroduces a modern version of what Lovejoy (1964) refers to as the "great chain of being," a metaphor used from medieval to modern times to express the constitutive hierarchical ordering of the universe. In its medieval form, the chain placed those things least like God at the bottom of a scheme of things that stretched from the "lowest" organisms, through plants and animals, to "higher" orders of perfection including humanity (who were made in God's image) and then degrees of angels. At the very top of the chain sat God as the most perfect being of all. This same model was later adapted in the eighteenth and nineteenth centuries to depict a biological and evolutionary order of developmental perfection that straightforwardly eradicated all those intangible orders above humanity. Mathews' own universal order is different only insofar as it replaces angels with ecosystems and religious or evolutionary principles with a "new world order" ultimately derived from the cybernetic prophecies of Norbert Weiner (1948), the sage of the information age. Here, in what is potentially the greatest irony of all for a deep ecologist,

Mathews bases intrinsic value on possession of the characteristics of an idealized computer. Of course, Mathews would argue that systems theory is based on principles that apply to animals as well as machines and that her hierarchy is qualified, as indeed it is. Indeed her whole thesis is ringed about with so many provisos and special considerations that even the casual observer begins to suspect her argument to be a rather contrived and convoluted way of reaching the conclusions she already had about what should and what should not count as having moral value. However, the real issue here is not Mathews' spurious argumentative strategy, nor her uncritical acceptance of a cybernetic metaphor that already carries too much weight in an age of information technology, but the manner in which she reconceptualizes both the subject and the object of ethics.[23]

Having begun with a forthright critique of the Newtonian mechanistic paradigm and atomistic conceptions of the self, she ends with an ethical system that apportions intrinsic value largely in terms of an individual's ability to possess self-interests, precisely the same feature that figures so large in anthropic and modernist accounts. In other words she grounds ethics in a conception of individual autonomy and self-regard that is almost identical to the possessive individualism that characterizes the most anthropocentric Enlightenment systems of ethics (Macpherson, 1979). Despite her radical pretensions there is little room here to give due consideration to others for their own sakes or for their differences from ourselves. The anchoring of ethical values in *self*-interest reflects, rather than challenges, the hegemony of that self-ish individualistic ideology that pervades modernity and capitalism. The fact that the individuals concerned are supposedly relative rather than absolute makes little difference, since it is only insofar as individuals are recognizably isolable systems that strive for their own persistence over time that they can be regarded as possessing intrinsic value. Similarly, the fact that subjecthood has been extended to a range of nonhuman "fuzzy" individuals is neither here nor there since others are regarded only in the light of their similarity to a unitary informational ideal. Those who would regard a rock or a waterfall as having any intrinsic value are, so far as Mathews is concerned, simply in error, for such things cannot be the objects of proper ethical consideration. They can enter the moral equation only insofar as they are of instrumental value for the maintenance of cybernetic systems or functional parts of those systems.

It must be clear by now that despite her biocentric conclusions Mathews' argument is, at root, another variety of "axiological extensionism." And, just as in all axiological extensionist arguments, she inevitably finds herself formalizing an ethical hierarchy that either fails to reflect our actual values or falls into an overexpansive egalitarianism.

This latter possibility emerges explicitly when, in order to temper her prima facie argument for an organizational hierarchy, Mathews suggests that because each system is nested in yet larger systems then the "differentials in the intrinsic values of different selves will 'average out,' and an effective egalitarianism in respect of intrinsic value ensue" (Mathews, 1993: 144). This may extract her from some messy problems that arise from arguing that complexity somehow equates with value, but does so only at the cost of raising new problems. For example, such a strategy implies that literally everything, from a cesspool to a squid and from a paramecium to a person must be regarded as equally valuable beings-in-themselves.[24] But to do so is not only impracticable, it is impossible. Such egalitarianism entails that every move we make becomes imbued with immense and overwhelming ethical significance, that every breath we take destroys innumerable entities of equal value to ourselves. Given these implications, it is not surprising that Mathews quickly abandons any consideration of the ethical prescriptions to be derived from her normative principles, despite the philosophical acrobatics she originally performed solely in order to justify them. She simply states that her principles suggest we should "'tread lightly' on this earth" (Mathews, 1993: 147).

Few who care about the state of our natural environment would want to argue with such a conclusion. But problems remain as to what Mathews means by "treading lightly" and why she thinks we should do so. Having all but abandoned the elaboration of any specific entailments of her ethical principles, Mathews turns instead to a metaphysical interpretation of "self-realization" that supposedly provides the motive for embarking on an ethical relation to nonhuman others. Humans, she says, in an unconsciously ironic reiteration of anthropic principles, are *special* because, unlike rocks or amoebas "they alone amongst these entities can *grasp* our unity with the greater whole" (Mathews, 1993: 149). As humans, we are by nature endowed with "self-interest, self-concern, self-love . . . [and when] we recognize the involvement of wider wholes in our identity, an expansion in the scope of our identity and hence in the scope of our self-love occurs" (Mathews, 1993: 149). In other words, Mathews reduces ethics an enlightened form of *self*-interest since our interest in others is limited to the extent in which we are able to recognize elements of our-selves in them. Nature's value to us is never anything more than an expression of (our own) self-interest on a grand scale. This is the very antithesis of giving an entity respect because of its being-in-itself.

Of course, Mathews is right to argue that things cannot be understood in themselves alone; that their intrinsic properties mean little or nothing by themselves. The solid mechanics of Newton's nature and Hobbes' society of isolated individuals will no longer suffice as a model for the complexities of either the natural or the social world in late modernity.[25] We now inhabit a world of relativity, where spaces are not empty gaps between concrete things, but are constitutive of the relational fabric of the world, a fabric that composes objects and subjects alike. But if this is true then the search for objective essences must cease and this includes cybernetic principles, like Mathews' *conatus,* which, having gained a certain tangibility in an age that reifies information, still operates like "sentience" or "rationality" to divide and rule the moral world.

Mathews is also right both to criticize the dominant anthropocentrism in ethical theory and to believe that the metaphysical system modernity employs to understand our place in the world affects the way we value it.[26] But these ideological effects are not straightforward and cannot be dispelled in a style of argument that adopts the both the logical form, and much of the ideological content, of the modern symbolic order. In other words I am not just concerned with the philosophical problems that arise in trying to apply extensionist theories. I am much more interested in drawing out and trying to account for the similarities in the form and content of the argumentation that Mathews' and Callicott's holism shares with the extensionism of Singer, Taylor, Attfield, and others.

What is clear is that, despite their protestations to the contrary, all of these supposedly biocentric systems remain indelibly anthropocentric, though this anthropocentrism is disguised in differing ways. Callicott and Mathews claims that we should treat nature with respect because it is part of our wider selves. But it is unclear why a system founded upon human egoism and self-interest should be regarded as having anything to do with ethics in a broader perspective. Why should I place myself at the center of things and regard my relations with my environment primarily in terms of my own desires for self-preservation and well being? This seems to have little in common with those ethics that promote an ideal of sympathy for/with others even at the cost of my own self-interest or the other's lack of similarity with myself.

For example, Iris Murdoch makes the point that, "[m]ore naturally, as well as more properly, we take a self-forgetful pleasure in the sheer alien pointless existence of animals, birds, stones and trees" (Murdoch, 1970: 85).[27] Nature offers us a chance to escape a world where all we see reflects "humanity" back at us, and perhaps the only long-term chance for the survival of nature lies in us coming to see it as being of value on its own terms. The appropriateness of using ethical language in discussions of environmental concerns lies not in the similarity of the moral objects to ourselves, but in morality's ability to express concerns about a wider community, a community not of formal equals or of a formal hierarchy but of complex interrelationships.

Singer's thesis of the expanding circle, and moral extensionism in general, is a graphic representation of anthropocentrism. Humanity sits at the center of a concentrically ordered nature, as the archetype of ethical value—both the measure and the measurer of all things. Though the extensionist rationale holds out the possibility of a formal egalitarianism, in practice, the greater the difference between us and them the less is the gravitational pull on our moral faculties.

In reality the continual discovery of new and better demarcation principles, of moral progress is a fiction, a "just so story" that echoes the modernist pretension of the onward march of (Western) civilization, of progress. The periphery of moral considerability is not determined by new scientific discoveries about our true relation to the world or by increasingly enlightened attitudes, it is determined by whatever feature or concept we are happiest with in any given historical and cultural circumstance. "Sentience," "flourishing," "having-a-good-of-one's-own," and "self-interest" are categories that appeal to a modern society mirroring as they do valued aspects of our own cultural background. For example, as Keith Tester has noted even the title of Singer's (1990) *Animal Liberation* operates to tie the question of animals' moral considerability "in with most of the other concerns about lifestyle and selfhood which appeared in the 1960's" (Tester, 1991: 168).[28] Singer seeks to make explicit links with those other liberation struggles that impinge on the popular consciousness of our time, thereby making his own proposals more acceptable. There is nothing wrong with trying to make such connections, but their existence should serve to remind us that philosophical rationales are not as autonomous as they claim to be.

My criticisms of extensionism should not be taken to mean that I believe that the categories utilized by these approaches to ground their arguments are entirely arbitrary; they are not. For example, our judging something sentient may, in certain circumstances, lead to us treating it differently—on a sunny day I see a worm "suffering" from dehydration on the road and in danger from passing traffic so I move it to the verge. In other circumstances it may make no difference whatsoever—I decide to dig the garden to plant some bulbs knowing full well that dozens of worms will be cut in two! But this very contextuality of both our values and the ethical categories the extensionist uses to operationalize our values, is precisely what is excluded by the so-called autonomy of reasoning. The only aspect of the situation that is supposed to be morally compelling is the question of whether or not worms are sentient. If they are, the act of gardening becomes a heinous crime. If they are not, saving the worm from immanent death becomes mere misplaced sentimentality.[29] The wider context—or, one might say, the *environment*—of our ethical values is precisely that which is excluded as immaterial in the stampede to provide rational grounds for morality. But this contextuality is not, as the extensionist myth would have us believe, a symptom of a messy, irrational, and uncivilized past in need of rationalization (in both senses of the term); rather, this disorderliness is actually the aspect of ethics that matters most.

Axiological extensionism is, I would suggest, ethically inappropriate, unworkable, and vastly oversimplistic.[30] Similarities of faculty are reified into universal demarcation principles in an attempted emulation of the natural sciences. The only empirical evidence admitted is scientific evidence on the distribution of the chosen demarcating faculty in the natural world. Thus intelligence testing, biological taxonomy, physics, ecology, and sociobiology are all admissible as evidence for the possible moral considerability of a class of objects. What is not admissible, though, is evidence about whether people actually do so regard an object. What is positively dismissed is the massive plurality of "reasons" why people can and do value things morally. The mania for objective theoretical criteria leads to a monolithic reductionism combined with an unwarranted mystification of one particular faculty as somehow bearing moral value.[31]

Perhaps the clearest way to see the problems this philosophical methodology creates is by looking at those things that are drawn out of moral bounds, things beyond the periphery of the expanded circle. In discussing his concept of the good of a being, Taylor contrasts a child with a pile of sand. The sand, he writes, has "no good of its own. It is not the sort of thing that can be included in the range of application of the concept entity-that-has-a-good-of-its-own" (Taylor, 1986: 61). This being so, it is excluded from moral considerability.

Yet we certainly can, and some of us do, extend moral considerability even to piles of sand in certain contexts (the context is all-important). The barchans, great crescent-shaped dunes found in the sand deserts of Arabia's "empty quarter" have inspired the imagination of many travelers and molded the lives of people like the Bedouin who have lived among them. The sandstone of regions like Exmoor or, more impressively, the Pakaraima mountains on the border of Guyana and Venezuela, containing some of the world's highest waterfalls, is directly responsible for their particular ambience. The feelings and forms of life these "piles of sand" have generated can and have led to their being valued for their own sake in ways that can best be described as ethical. To take a different example, when oil spills from tankers onto sandy beaches we think such avoidable occurrences morally reprehensible, not just because they are aesthetically displeasing but because it makes sense to talk of *desecration* of the beaches.

In excluding the contextual aspects of ethical evaluation, by placing social and environmental contexts outside of that which properly constitutes the ethical, these philosophical methodologies reconstitute and rationalize the moral field. This rationalization is, I shall argue, a principal characteristic of modernity and modernism. Indeed the most significant link between all of the methodologies for allocating values to nature, whether they are economic, axiological, or holistic/individualistic, is their tacit and unspoken agreement about the kind of argument and approach to valuation that is acceptable and necessary. Apparent differences of opinion mask a deeper and more fundamental conformity within contemporary philosophy. One way of characterizing this conformity is in terms of their acceptance and embodiment of what has been termed "formal rationality" and it is to this I now turn.

Rationalization and the Context of Modernity

Up until now I have used the terms "rationalization" and "formal rationality" without making any specific reference to their sociotheoretical origins. I want now to make amends for this and try to show exactly why my characterization of current moral problematics in terms of the rationalization process might be relevant to radical environmental critiques.

Rationalization is, without doubt, the key concept in Max Weber's sociology of modern Western society. As one of the founding figures of sociology, Weber was concerned to identify and understand the distinguishing features of modernity, that is, of a social system that, unlike its historical predecessors and geographical contemporaries, seemed to operate efficiently despite the fragmentation of its social practices and moral values. He believed that modernity's uniqueness lay in "the 'specific and peculiar rationalization' that distinguishes modern Western civilization from every other" (Brubaker, 1984: 1). In this sense Weber is the sociologist of modernity par excellence, as a number of recent texts have testified.[32]

The complexities and nuances of Weber's multifaceted conception of rationalization are well documented (Kahlberg, 1980: 1145–79). But, for our present purposes it suffices to say that rationalization is associated with the loss of tightly knit communities, the decline of religion, and the rise of capitalism. In Western society these trends went hand in glove with the rise of specialized fields of knowledge, especially the sciences, and the extension of scientific methodologies and worldviews.

The fragmentation of social practices that accompanies the increasing complexity of modern society reduces the possibility for shared values within society as a whole. Values formed by the individual's immersion in innumerable separate practices become increasingly incommensurable.

People with different social backgrounds will occupy different "value spheres" and therefore often see each other's actions as irrational. For Weber, these value-laden backgrounds, acquired through the individual's membership of particular social practices, are not open to judgment by reason but form the unquestioned bedrock of our values. In Stephen Kahlberg's words, "[t]hese value constellations, even though for Weber they are themselves largely manifestations of *irrational* historical, economic, political, and even geographical forces . . . constitute rationally consistent world views to which individuals may orient their action in all spheres of life" (Kahlberg, 1980: 1170). Thus, as Rogers Brubaker suggests, "[u]nderlying Weber's emphasis on the limits of rationality is the idea that irreconcilable value conflict is inevitable in the modern social world" (Brubaker, 1984: 5).

Ironically the only shared assumptions left for rationality to work upon that can command agreement are those of the alienated experience of individualism that this fragmentation of society gives rise to. In this increasingly individualistic society lacking any overarching moral principles, people inevitably come, via self-reflection, to act on the basis of instrumentality, treating things as means to their own particular ends whatever they might be; this Weber terms "subjective rationality." Such rationality may be fine for the personal sphere, but there remains a desperate need to facilitate social exchange in an increasingly depersonalized public sphere. This void can only be filled by a particular form of reason that gains currency precisely because it does not "take sides" in conflicts between value spheres and claims to be neutral with respect to individuals' particular ends. This Weber refers to as "formal rationality."

Formal rationality takes as its model those aspects of the new order that seem most successful. Foremost amongst these are the development of capitalism with its associated economic rationale and monetary calculus, the burgeoning field of mathematics, and abstract theories in the natural sciences. (These developments in business and technology had, after all, been of prime import in bringing about a consequent diversification of society, which led to the alienation of individuals from their traditional cultures and values.) Formal rationality emphasizes the need to subject all areas of human relationships to a logic of explicit rules and calculative efficiency.[33] In placing its emphasis on means rather than ends it can avoid the charge of partiality and claims to order society according to neutral rational principles. In this is it distinguished from what Weber refers to as a "substantive rationality" that always includes particular value commitments.

All that was necessary to develop modernity in its present form was a reorientation of the individuals composing society, so as to accept the right of formal rationality to govern their lives. Weber saw the "this worldly" and self-reflexive aspects of the Protestantism that developed in Europe from the sixteenth century onwards as the necessary catalyst for the internalization of this process of rationalization. "Science, and modern rationality more generally, represents the Puritan obsession with calculation, impersonal rules and self-discipline without the Puritan belief in their divine origin. It is Puritan epistemology without Puritan ontology" (Alexander, 1987: 191).

As the rate of change and fragmentation grew so the pressure to extend these methodologies and their accompanying ideology to other spheres of society became overwhelming. Rationalization took different forms in different spheres of life. Juridically it appeared as formalized and explicit laws and penalties in the increasingly rigid bureaucratic structure of the legal system. In industry it took the form of systematic accounting procedures and of administrative procedures designed to increase efficiency (of which the time and motion studies promoted by Taylor and the techniques of "human resource management" might be prime examples). In politics charismatic leadership was replaced by party structures and bureaucracy. Rationalization's effects began to be felt in every sphere, even, Weber claims, including the introduction of formal notation systems in music. Human activity in all its varied expressions became more and more regulated, codified, and formalized in explicit rules and laws, and subject to criteria of calculability. As this rationalization spread it also brings about the disenchantment (*Entzauberung*) of the world.[34] Those things that had previously been value-laden and mysterious came to lose their inherent value and be seen only as means to particular ends.[35]

Of course, rationalization goes hand in glove with the emergence of the typically modern and Western form of bureaucratic administration and government. Formal rationality is ideally suited to such purposes since it involves the systemic application of (supposedly) universal methods that are deemed capable of resolving value conflicts via codified rules or abstract calculations. Formal rationality and bureaucratic and economic structures thus increasingly dominate Western society. Where substantive rationality recognized that values and background beliefs were important and integral parts of rational choice, the logic of modernity assumes that value conflict can be resolved only by recourse to formal rationality. This "absolutely inescapable condition of our entire [modern] existence" (Weber in Marcuse, 1968: 204) is what Weber referred to as the "iron cage" of capitalism.

In this way Weber is the first social theorist to give a detailed account of the relationship between the development of a subjective rationality, that is, the conscious reflection on one's individual aims, with a social and formal rationality, the supra-individual processes by which particular types of explicit rational structures ramify throughout the social arena. He links together two of the central features of modernity: its conception of the subject as autonomous and self-interested, and the increasing role played by formal reason in the public sphere. Both are particular instances of the process of rationalization. "By emphasizing the historical connection between new forms of institutionalized control over men and a new ethos of self-control, between institutionalized discipline and self-discipline, Weber supplements institutional with psychological analysis in an effort to clarify the relation between social structure and personality" (Brubaker, 1984: 35).[36] In summary, according to Weber, modern Western society is peculiar in the fact that it is composed of isolated individuals who, while feeling themselves to be autonomous have actually become dominated by formal rationality. This rationality has no communal substantive content but comes to regulate the life of its citizens with its own internal logic of enhancing efficiency, calculability, and codification.

The Frankfurt School and Instrumental Rationality

Despite Weber's strictures about the "iron cage" of capitalism and his "metaphysical pathos" over the increasing dominance of bureaucracies, he retained something of an ambivalent attitude toward the process of rationalization as a whole. He thought that it had brought undoubted benefits, not just in terms of scientific advancement or social organization, but in terms of the greater freedoms enjoyed by the majority of those living in systems dominated by self-reflexive rather than hierarchical religious worldviews.[37] On the other hand the growth of formal rationality now threatened the individual with new forms of domination and oppression by ordering her life according to rules of efficiency rather than according to her needs. The same rationality that depends upon, and to some extent facilitates individuality as we now recognize it, also, paradoxically, threatens its demise as substantive rationality becomes marginalized in the public sphere.

The claims of formal rationality to be value-free seem increasingly hollow. Its insistence on the universal application of abstract calculations and the abstract codification of laws inevitably distort the values actually held by individuals and communities. We become faced with a cult of efficiency for its own sake, the tyranny of quantification in all areas of life, an increasingly intrusive juridical system, and the reification of a form of reason that, while claiming to serve everyone's interests, actually suits few but bureaucrats and administrators. (Formal [instrumental] rationality and efficiency are often appealed to in order to justify the manager's right to manage, that modern equivalent of the divine right of kings.) The individual everywhere becomes subject to new forms of regulation and control, all justified by the democratic pretensions of formal rationality.

Weber's ambivalence about formal rationality turns into a passionate repudiation of its totalitarian aspects by the Frankfurt School, a group of Marxist social theorists grouped around the Institute for Social Research.[38] These writers, including Max Horkheimer, Theodor Adorno, and Herbert Marcuse, convinced of the power of formal rationality in its unholy alliance with capitalism and the mass media, emphasize the import of technology in fostering the rationalization process and facilitating social regulation. In his *Eclipse of Reason* Horkheimer (1974) traces that same process of the formalization of reason, which he terms "subjective reason," in modern society.[39]

Marcuse famously regards science, technology, and the formal (or in his terms "instrumental") rationality that accompany them as producing a society with only "one dimension"—a society where it becomes impossible even to think outside of the limits imposed by these rationales. We accept without question the current order of things since scientific techniques and capitalism continue to come up with the material goods to satisfy the desires that the complimentary techniques of the mass media, particularly advertising, constantly create. In Marcuse's words, "[a] comfortable, smooth, reasonable, democratic unfreedom prevails in advanced industrial civilization, a token of technical progress" (Marcuse, 1991: 1). This Orwellian scenario seemingly makes even "thought-crimes" an impossibility since modern society excludes its citizens from the possibility of questioning its aims, methods, or values.[40]

Perhaps now we are in a position to see the direct relevance of Weber and the Frankfurt School's analysis of formal (instrumental) rationality for environmental ethics. First, because it gives us an account of what happens to values in the context of modernity. Second, because it explains the predominance of economic and instrumental valuations in modernity. Third, because it can also give an account of the formal axiological pretensions of extensionist ethical theories. Fourth, because it helps make explicit links between the dominance of formal rationality and the maltreatment of the environment. I have, perhaps, said enough about the first and second issues to make my point, but want to conclude this chapter by drawing together the issues of the philosophical ethics of the environment and our current environmental predicament.

Weber and Frankfurt School thinkers like Horkheimer and Marcuse, argue that formal (instrumental) rationality in alliance with technology becomes an end in itself. In the quest for increased efficiency and material gain, modernity reduces everything to a means without any foreseeable end and feeds its hungry furnaces with fuel composed of the surrounding natural environment. For nature, now thoroughly disenchanted and stripped of all inherent or substantive value becomes, like everything else, nothing more than a resource subject to processes of calculation.[41] Marcuse was perhaps foremost in making explicit connections between the process of rationalization and the evaluation of nature. "The basis of this rationality is abstraction which, at once theoretical and practical, the work of both scientific and social organization, determines the capitalist period. . . . As universal functionalization (which finds its economic expression in exchange value) it becomes the precondition of calculable efficiency. . . . Abstract reason becomes concrete in the calculable and calculated *domination* of nature and man" (Marcuse 1968: 205).[42]

Even critics of the Frankfurt School like Perry Anderson have to agree that they were ahead of their time in recognizing and opposing the results of rationalization in terms of the destruction of the natural environment. Indeed, he states: "Adorno and Horkheimer called into question the very idea of man's ultimate mastery of nature" (Anderson, 1984: 89). The rationalization that characterizes modernity entirely changes the relationship between culture and nature. As Martin Jay puts it: "At the root of the Enlightenment's program of domination, Horkheimer and Adorno charged, was a secularized version of the religious belief that God controlled the world. As a result, the human subject confronted the natural object as an inferior, external other. At least primitive animism, for all its lack of self-consciousness, had expressed an awareness of the interpenetration of the two spheres [the human and the natural]. This was totally lost in Enlightenment thought, where the world was seen as composed of lifeless, fungible atoms" (Jay, 1973: 260). Horkheimer neatly sums up this process: "Economic and social forces take on the character of blind natural powers that man, in order to preserve himself, must dominate by adjusting himself to them. As the end result of this process, we have on the one hand the self, the absolute ego emptied of all substance except its attempt to transform everything in heaven and earth into means for its preservation, and on the other hand an empty nature degraded to mere material, mere stuff to be dominated" (Horkheimer, 1974: 97).

If the prevalence of formal rationality can help explain modernity's destructive relationship with the natural world, its pervasiveness should also alert us to the possibility of its presence in our theoretical accounts. My earlier claim was that axiological theories might be regarded as products of their specifically "modern" time and place. This claim can now be filled out in terms of their embodiment of the principle of formal rationality. Indeed they should be seen as exemplars of the manner in which rationalization has taken hold even in the philosophical field.[43]

The ethical theories mentioned earlier in this chapter can be seen as uncritical incorporations of two differing strands of formal rationality. The utilitarian approach advocated by Singer is, as previously mentioned, simply the moral equivalent of the economization and quantification of values. It precisely mirrors the accounting procedures of the economic sphere, engaging in a "hedonistic calculus" that claims to impartially sum the pros and cons of any given situation on the basis of a formal individual equality. Alternatively, those who attempt to allocate moral standing, perhaps even rights, to elements of the natural environment, also make much of the formal equality given to all who fit certain pregiven requirements; sentience, having-a-good-of-one's-own, and so forth. Their approach has more in common with the rationalization of the juridical sphere noted by Weber in that it sees the answer to moral dilemmas as lying in our identification of explicit abstract criteria that are then utilized to make impartial decisions. In Weber's terms such "legal authority" rests "on a belief in the 'legality' of patterns of normative rules" (Weber, 1964: 328); together these compose a "consistent system of abstract rules which have normally been intentionally established" (Weber, 1964: 330). Here then the emphasis is on a formal equality under the rule of law(s).

What matters most in this ethical modernism is that an order is established whether in terms of a uniform calculable metric or an explicit and internally consistent set of codes. In neither case can a "single action . . . stand by itself or be valued on its own merits alone, but only in terms of its bearing on a whole system of rational conduct" (Parsons in Weber, 1964: 80). So defined, ethics turns in on itself, reformulating the boundaries and content of the moral field in terms of formally "rational conduct."[44] All moral action has to be explained and justified in terms of formal rationality. This is, of course, exactly what Singer and colleagues regard as the "autonomy of reasoning" but this autonomy is not all that it seems and it comes at a price.

All explanations and justifications of moral behavior must be rationalized, translated into terms that are capable of having currency within formal systems. Modernist ethics then reifies these reasons, regarding them as the material, efficient, formal, and final causes of moral behavior.[45] But as Weber points out, they are not. Our values emerge from our relations to the social (and I will argue natural) environments in which we find ourselves, that is, they are *substantive*. This environment, and our relation to it, is the material ground of ethics and any explanations or justifications of ethics must appeal to this substantive background, to our "form(s) of life." While our particular moral stances may have been affected by moral arguments (since rationalization is part of the substantive background of modernity), these are only rarely (if ever) the immediate cause of our ethics. They are not then the efficient cause of behavior that is usually engaged in habitually, or according to our current feelings. Nor do we usually engage in moral action for a reason, we do so for some-thing and any reasons given are generally post hoc rationalizations of ethical values and activities. It is these rationalizations that formal ethics seizes upon as the be-all and end-all of morality. Anything that cannot be translated into this formal reconfiguration of the ethical field is regarded as immaterial or irrational. In other words, despite their claims, such formal philosophical arguments are not autonomous or value-free but align themselves with the process of rationalization. They are only autonomous in the sense that they close themselves off from taking substantive, sociological, and natural contexts and explanations of values into account.[46]

Where economics is at least explicit in its aims and methods, axiological extensionism and extensionist varieties of holism have failed to recognize that the motivation for, and the effects of, their ethical methodologies are identical with such rationalization. They too seek to institute (and often institutionalize) formal acontextual systems that allow the trade-off of values in order to determine the "right" course of action.[47] Extensionist claims to embody objective facts and apply a neutral rationality both echo and support the rationale of those who profess to be mere disinterested administrators in modernity's legal/bureaucratic and economic structures. While these administrators claim to be simply *applying* neutral and universal frameworks to particular cases it is precisely their supposed "disinterestedness," their lack of involvement in messy contexts, which serves as the justification for them making abstract evaluative decisions for others.

In short, the dominant forms of moral theory have much in common with neoclassical economics in attempting to provide an acontextual rubric that reduces ethics to a theoretical tool for passing judgments or evaluating actions. Such formal rubrics facilitate managerial and technical efficiency. They promulgate and support the myth of a neutral rationality in the hands of professed experts as an impartial arbiter in the service of society as a whole.[48]

The supposed impartiality of current moral axiologies, and therefore the kinds of extensionism that seek to extend these axiologies, is false. They actually rely upon and serve to extend that very principle of rationalization that underlies modernity, that is, that same principle that is, if Weber and the Frankfurt School are correct, also implicated in our irresponsible attitude to the environment in the first place. This indeed is a case where Alisdair MacIntyre's warning is apt, where "what we take to be resources for the treatment of our condition may turn out themselves to be infected areas" (MacIntyre, 1981: 3–4).

Concluding Remarks

The moral considerability of nature need not be a matter of discovering abstract criteria by which one can judge such valuations right or wrong in any absolute sense. Rather, ethical values need to be explained and justified in terms of their context and origins, their production and their reproduction in particular social and environmental circumstances. This contextualization is, I shall argue, exactly what radical environmental approaches need and must support. One of its first priorities is to understand where modernist philosophical ethics is coming from in order to account for its successes and failures in capturing our ethical intuitions about our relations with our environment. Such contextualization requires that philosophy becomes reflexive—throws over its grandiose claims to universal significance and accounts for its "stories" in terms other than their being the logical working out of disinterested reason. It requires a theorization of society and of society's place in historical and natural environments, a theorization that is aware of its own embeddedness in time and space.

Our present moral field has become entrapped and constrained within theoretical paradigms that, despite their supposed differences, actually agree in all fundamental ways on what form ethics, and hence environmental ethics, must take.[49] This form (and formalism) should be regarded as antithetical to the aims of radical environmentalists. Our reasons for valuing the jungle or the primrose, the desert or Antarctica are manifold. Not only are they often concerned more with our perception of their disparity from humanity than any affinities, whether natural or intellectually contrived, but the whole enterprise of providing necessary criteria, or grounds for moral considerability reach the end of the road where environmentalism begins. Such methodologies serve only to impose too rigid a structure upon our moral beliefs and values. If moral consideration is to be extended to nonhumans, this has to be done not on the spurious basis of shared properties but on due recognition of other natural phenomena for their differences as well as their similarities and the many and varied ways we can relate to them.

The predominance of formal frameworks has until recently ensured that genuinely oppositional moral values and narratives that refuse to be reduced to an abstract formalism are excluded from the public realm.[50] For this reason it is important that radical environmentalism does not fall victim to the interminable maze of moral argument that currently passes as the sum total of the moral sphere. Those moral theories, like utilitarianism, that have a practicable intelligibility in terms of a reductive efficiency in modern bureaucratic society would lose this entirely in a society reconstituted and restructured on radically different lines. In this sense utilizing current moral paradigms may actually be counterproductive insofar as reaching any radically different form of social and environmental relations is concerned. In doing so we might simply reinforce a form of philosophy that is complicit in the incorporation of environmental values into the present social structure, rather than supporting social change.

2

Closed to Nature: Social Theory and Ethics from Durkheim to Habermas

The previous chapter suggested that, generally speaking, philosophical approaches to environmental values have failed to recognize the degree or the relevance of their own entanglement within the temporal, spatial, and social horizons of modernity. Their search for universal truths, for methods or criteria that will stand the test of time, is symptomatic of a style of philosophizing that pays little heed to its own situation. This lack of reflexivity is especially noticeable in those approaches that define themselves as "analytic"; a school of thought that, beginning in the early twentieth century at Cambridge and Oxford, has expanded to exercise an almost complete hegemony in many universities, especially in Britain (Rée, 1993; Smith, 1998). It was this (analytic) approach that Marcuse identified as exemplifying one-dimensional thought, as a mode of thinking that both embodies and furthers the interests and concerns of the current order of things. The axiological approaches to ethics mentioned in chapter 1 could all, to greater or lesser degrees, be regarded as falling within this dominant analytic paradigm. Insofar as they comply with present realities they fail to undertake what Marcuse considered to be philosophy's primary tasks; the recognition of "the limits and the deceptions of the prevailing rationality" (Marcuse, 1991: 186) and "the intellectual dissolution and even subversion of the given facts" (Marcuse, 1991: 185).

Dumped by the historical wayside in seventeenth- and eighteenth-century Europe as science, flushed with its practical successes, drove off alone and at speed toward an as yet unspecifiable future, philosophy picked itself up, dusted itself down and reconfigured itself in the early twentieth century as analysis. In this way it hoped to follow in science's tracks and perhaps even hitch the occasional ride on the scientific juggernaut. (Occasionally, suffering from a kind of retrograde amnesia, it even imagines itself to be back in the driving seat!) Contemporary analytic philosophy conceives itself as having a symbiotic relationship with science due to their common application of abstract reasoning to the given facts. In Anthony Quinton's words, the analytic approach is "respectful of science, both as a paradigm of reasonable belief and in conformity with its argumentative rigor, its clarity and its determination to be objective" (Quinton, 1995: 30).

But this desire to emulate the natural sciences' supposed objectivity, comes at the cost of failing to recognize the relevance of social and historical influences. As Bell and Cooper admit in the introduction to their collection on the analytic tradition, "analytic philosophy has been, and remains, largely unselfconscious and almost entirely unhistorical" (Bell & Cooper, 1990: vi).[1] Analytic philosophy is notorious for removing arguments from their historical setting and passing judgement upon them as good or bad, coherent or incoherent, logical or illogical. In abstracting arguments from their original sociohistorical nexus analysts set themselves up as disinterested inquisitors, as the very embodiment of a neutral and timeless rationality. Thus, for example, a utilitarian core is extracted from Bentham's works and examined in philosophical isolation for its merits and demerits. Any question about the origins of utilitarianism, in terms of its relation to the French Revolution, the industrialization of society, the development of bureaucracies, and so on is deemed the role of the historian or sociologist and not that of the philosopher. Yet this is precisely to miss a series of vital and interesting questions about how a theory like utilitarianism might relate to the structure of a particular society and hence how it succeeds or fails in expressing the moral intuitions that have currency within that society.

This philosophical problem thus has two interlocking components. The refusal to recognize the influence of historical, social, and environmental factors on theory formation goes hand in glove with a disciplinary parochialism. That is to say, an insistence on the disciplinary distinctiveness of (analytic) philosophy has led to the construction of impervious boundaries delimiting what does and what does not count as of philosophical relevance.[2] For example, Anthony Flew informs us in the introduction to his *Dictionary of Philosophy* that "Plato is asking the philosopher's logical, conceptual and semantic questions and these are altogether different from the factual questions asked by the psychologist or the physiologist" (Flew, 1979: xiii). Having separated philosophy from empirical concerns Flew also explains why Western philosophers need have no truck with the theoretical canon of different cultures. It is, he says, lack of genuine argument "not European parochialism [. . . that explains] why the classics of Chinese philosophy get short shrift. The Analects of Confucius and the Book of Mencius are both splendid, *of their kind*. But neither sage shows much interest in the sort of question thrashed out in the Thaetetus" (Flew, 1979: ix, emphasis added).

Flew also differentiates philosophical questions from questions about the sociology of knowledge. This handy distinction specifically rules out "asking what social conditions promote or inhibit the acquisition of what sorts of knowledge" (Flew, 1979: viii). The philosopher can thus inhabit a discursive field of pure thought without any need to be reflexive about their relations to society at large. We are presented with a picture where the disinterested philosophical mind reflects the world in the faultless mirror of its rational thinking, where concepts are cleaned and polished and returned gleaming new to a grateful populace. Yet from another point of view this very conception of philosophy expresses and enshrines the institutionalized intellectual elitism of the Oxford common room.[3] What is more, the (recent) development of these disciplinary boundaries could itself be regarded as one more example of the theoretical effects of the increasing division of labor and the fragmentation of perspectives and values that are so characteristic of modernity. Paradoxically philosophy's determination to defend its (new-found) integrity and proclaim the unsullied purity of its reasoning might itself be regarded as symptomatic of its situation within a particular set of social and historical circumstances.

The last chapter sought to dispel any notion that philosophical argument could be as autonomous as these analytic philosophers seem to suggest. It argued that philosophy cannot afford to ignore its dependency upon and relations to its surrounding social (and natural) environment. The sociotheoretical analyses of Weber and the Frankfurt School were used to help uncover some of the largely unrecognized links between the form and content of moral axiologies and the pervasive influence of the rationalization process in modern society. Given these arguments it might seem reasonable to abandon philosophy to its myopic predilections and look instead to social theory for an exposition of the nature of ethics and a potential ethics of nature. However, not all philosophies are so short-sighted or so subject to disciplinary strictures, and unfortunately, as these next two chapters will suggest, much mainstream social theory is currently in little better shape than philosophy to offer assistance to radical environmentalism.

The reasons for this impasse are similar to those that dog philosophy. Certain social theoretical paradigms, most especially those that like analytic philosophy ally themselves with a scientific approach, have tried to justify their disciplinary distinctiveness in terms of the existence of a right and proper object of their investigations, namely society.[4] The concept of society, thus reified, comes to delimit not only the boundaries of society as *explicandum* (that to be explained) but also sociology as *explicans* (that which is called upon to give an explanation). To retain their disciplinary distinctiveness, sociology, and social theory must explicate their subject matter in terms of specifically *social* causes or processes—rather than, for example, by utilizing biological, psychological, or philosophical explanations. Just as analytic philosophy delimits the ethical in its own terms as something to be defined by and within a field of abstract reasoning, so sociology too places its own boundaries on the form that ethics can legitimately take.

From the point of view of radical ecology the result is that the moral field again finds itself reconfigured in order to fit theoretical forms that exclude their concerns. First, as we shall see, following Emile Durkheim, social theory has offered a predominantly normative and functional characterization of ethics as a field of prescriptive forces that operate as a kind of social glue, resisting change, and fixing current social relations in place. This characterization is unfortunate since it necessarily neglects the ethical critiques of current social relations and the ethical arguments for social change favored by radical ecologists. Second, sociology as a discipline defines itself in terms that accept and exemplify that nature/culture dichotomy that is such a prominent feature of modernity; a dichotomy that radical ecology seeks to question (e.g., Plumwood, 1993). The disciplinary origins of much social theory thus place limits on its ability to take environmental issues seriously.[5]

Since Durkheim plays a key role in instituting sociology as a discipline and formulating social theory's dominant approach to ethics I begin with an account of his writings. I then follow this normative account of ethics through the functionalist and cybernetic approaches of the mid-twentieth century, typified by the work of Talcott Parsons, to the contemporary neofunctionalism of Jürgen Habermas.[6] This is not intended to be a comprehensive history of social theoretical engagements with ethics but to give an insight into the treatment of ethics by those theoretical approaches that, however consciously or unconsciously, have allied themselves with the process of rationalization and with modernism.

There is a certain irony that it should be this same functionalist strand of social theory that has, almost uniquely, regarded ethics as a vital and integral locus for any social analysis. But, for this very reason, and despite the problems raised below, I believe it may be worthwhile to try and recuperate something from these approaches. I thus conclude the chapter by outlining an alternative contextual reading of Durkheim. This reading tries to bring out an incipient spatial metaphorics (made explicit in his term "social morphology") latent in Durkheim's theoretical framework. I thereby hope to move some way toward the antifoundationalist "ethics of place"—the environmental ethos—that I seek to develop in later chapters. However, in order to facilitate this reading it is necessary to deconstruct some of the key modernist assumptions underlying Durkheim's theoretical stance. I shall therefore challenge the totalizing nature of his normative account of ethics, the anthropic and disciplinary boundaries of his methodological assumptions, and his linear account of social evolution and progress.

Durkheim and Ethics

Social theory has tended to regard Durkheim's emphasis on morality as a somewhat quaint and outdated example of fin-de-siècle anthropological concerns with "primitive" religiosity. Indeed, for Durkheim, ethics and religion were originally one—they occupied the same social space. "It is no more right to conceive of religion as emerging from morality than morality from religion. In the beginning, law, morality and religion were combined in a synthesis from which it is impossible to dissociate the elements" (Durkheim, 1993: 92). The "primitive" social group was, Durkheim thought, composed of relatively homogenous individuals closely tied to each other through a *conscience collective* and collective representations of religious significance.[7] This conscience collective was at one and the same time a common *consciousness* in the sense of being a shared worldview and a common *conscience* in the sense of a moral ground for collective values and practices. Durkheim referred to this kind of primitive consensual association as exhibiting a "mechanical solidarity."

The conscience collective is a social phenomenon that transcends, even though it is embodied in, the conscience and consciousness of individuals. This is made explicit in Durkheim's famous dichotomy between the sacred and the profane. The profane encompasses those aspects of society that are the remit of the individual qua individual. Things within the profane sphere can be transformed or appropriated by individuals without invoking sanctions from society as a whole. The sacred by contrast is that which transcends the wants or needs of the individual—it is beyond appropriation. In Durkheim's view this is precisely because it is a product and embodiment of the social formation itself. The sacred both incorporates and symbolically expresses the given social order, thereby helping to maintain and transmit that order through time and across generations. Thus, to behave sacrilegiously is to transgress and challenge the social order itself.

The supra-individual authority wielded by the conscience collective determines what is and is not regarded as sacred, but the ethical and the sacred are not just "elementary" features of "primitive" societies. As is well known, Durkheim employs a linear "progressive" model of social evolution. This evolution from primitive to modern is marked by the increasing division of labor. The consequent fragmentation of unified social practices and worldviews leads to the dissolution of mechanical solidarity and its gradual replacement by a system dominated by "organic solidarity," that is, the mutual interdependence of structurally differentiated social groups engaged in specialist tasks but each (like the organs of the body) requiring the other to survive. As each group gains relative autonomy in terms of its social practices, its worldviews, its interests, and its norms diverge, thus modern organic society is a system necessitating greater dependence on others and more communication with them, yet one where, ironically, the conscience collective can no longer predominate.

In organic solidarity collective representations are much reduced and where they remain they do so only by becoming less specific and more abstract. Individualism itself now becomes a ground of organic solidarity—a vague ideal that can serve to unite people if only by emphasizing the differences that delineate their particularity. Society is secularized, as the transcendental space the sacred occupied mutates into the worship of those individual interests that were previously the hallmark of profanity. In Durkheim's words: "As all the other beliefs and all the other practices take on an increasingly less religious character, the individual becomes the object of a sort of religion. We have a cult of personal dignity" (Durkheim in Lukes, 1988: 157).

This cult of individualism still derives its moral force from society though its social origins remain largely unrecognized. (Just as the social origins of the sacred were hidden from individuals in primitive society.) To the casual eye, this rampant individualism might imply the end of ethics—its dissolution into an egoistic autophilia. At the very least it would seem to generate an inherently unstable situation—an individualism guaranteed and reproduced socially, yet an individualism that by its very nature is unlikely to credit society with any transcendent properties.[8] Yet, for Durkheim, an ethic of individualism meant respect for all individuals, for their dignity and respective roles, for their differences. The cult of the abstract individual ensured that each particular individual signified something sacred.

Durkheim and the Sociological Closure of the Moral Field

Durkheim ingeniously marries together an account of the nature of ethics and society (an ontology) and a methodological positivism (an epistemology). He provides an account of society and sociology. As Lukes, quoting Durkheim, states "for sociology to be possible 'it must above all have an object of its own' a 'reality which is not in the domain of the other sciences'" (Lukes, 1988: 9). It is hardly going too far to suggest that morality is *the* archetypal sociological object, a social phenomenon to be understood in terms of *the* science of social causation. As Lukes points out, society is the *fons et origo* of morality—its "causal determinant, cognitive and symbolic referent and functional consequence" (Lukes, 1988: 481). This closure at the level of society (ontology) is supported by and reflected in the (epistemological) closure of post-Durkheimian sociology as a disciplinary field that allows only social explanations of social phenomena.

In spatial terms, Durkheim posits both an autonomous bounded social space and an autonomous and bounded science whose job it is to investigate and account for that space. Each enclosed system theoretically supports and justifies the other. To broach the wall between the social and the asocial would also to be to undermine the autonomy of sociology as a discipline. Similarly, to allow for extradisciplinary involvement would be to suggest the permeability of the social, to undermine its "objective" status.

This closure can be seen in Durkheim's rejection of animistic and naturalistic accounts of the origins of "elementary" religiosity (Durkheim, 1968). Animism, favored by such figures as Sir E. B. Tylor and Herbert Spencer, focused on the religious fascination with the existence of a nebulous and often insensible other world of spirits, souls, demons, and so forth existing behind or in parallel to the sensible everyday world. They suggested that these animistic concepts must come from misunderstandings in our experiences of different psychological states such as the relation between waking and dreaming. Yet Durkheim dismisses this explanation on the grounds that this makes religion (a major social phenomenon) nothing more than a "tissue of illusion". "What sort of a science is it" he asks "whose principal discovery is that the subject of which it treats does not exist" (Durkheim, 1968: 70).

The naturalists, like Max Müller, set out to explain the power and authority of religion as emanating form the awe inspired by nature's spectacular displays, like thunder, lightning, or the seemingly inexhaustible torrents of mountain streams. Durkheim rejects Müller's view that "[t]here is no aspect of nature which is not fitted to awaken within us this overwhelming sensation of an infinity which surrounds and dominates us" (Durkheim, 1968: 74). For Durkheim this is to try to explain the sacred (and social) in terms of the profane (and natural). The sacred has special qualities that need a qualitatively different explanatory cause:

> Nature is always and everywhere of the same sort. It matters little that it extends to infinity. . . . The space which I imagine beyond the horizon is still space identical with that which I see. The time which flows without end is

> made up of moments identical with those I have passed through. Extension, like duration, repeats itself indefinitely; if the portions which I touch have of themselves no sacred character, where did the others get theirs? . . . A world of profane things may be unlimited; but it remains a profane world. (Durkheim, 1968: 85)

Thus Durkheim dismisses out of hand both naturalism and animism on the grounds that they do not fit his methodological criteria, they try to explain a social phenomenon by recourse to things (whether psychological or natural) that fall outside of social reality. The inevitable results of Durkheim's strategy are that the religious and ethical are severed from their extrasocial settings and that society and sociology become closed to nature.

Ironically, support for this ontological and epistemological closure remains just as vehement today among sociologists who are far from positivistic in other ways. Thus the accusations of biological determinism that meet any talk of the natural construction of society can only really be understood in terms of defending disciplinary boundaries, by emphasizing, just as Durkheim did, the need to explain social phenomenon wholly in terms of social causation.[9] Chapter 4 will consider this problem in detail with respect to explanations of environmentalism in terms of the *social* construction of nature, that is, its social appropriation in discourse, symbolic fields, regimes of power, and so forth. However, I want to turn now to the profound consequences this ontological and epistemological closure has for ethics. In particular, how the admission of the ethical to the pantheon of sociological objects, that is, of social facts, makes a normative redefinition of ethics almost inevitable.

The Sociotheoretical Transformation of the Moral Field: Durkheim and Ethical Normativism

Ethics has all the features that Durkheim's positivistic approach requires to make it a "social fact," that is, an object amenable to sociological study. It is *external*—existing before the individual's birth and after their death in a shared social space. It is *general*—its pervasive presence is felt and can be observed throughout that society. It is *constraining*—that is, it governs the particular behavior of individuals, regulating their actions and promoting praise or sanctions to reinforce its normative force.

For Durkheim ethics is normative in two closely related senses. First, ethical values are not abstract absolutes but variable norms embodied in specific social formations. Durkheim's methodological positivism claims to eschew meta-ethical philosophical debates that attempt to define "the ethical" in isolation from its existing social forms. For Durkheim, all discussion of moral life should begin by studying diverse types of moral conduct in the specific contexts of particular societies. Abstract debates about the nature of right or wrong per se are dismissed in favor of a socially relativistic analysis of compliance or noncompliance with existing social norms.[10] This methodological contextualism is both possible and necessary because ethical norms are *external* and *general* in any given society.

Second, morality is normative in the sense of being prescriptive—circumscribing the boundaries of socially acceptable behavior and encouraging conformity. Ethics has both a regulative and a reproductive role, functioning to ensure social cohesion and continuation around specific norms. This is so whether or not the norms are made explicit in rules and laws or are implicit in the way in which certain behaviors accrue praise or blame. This is also true whether the laws concerned are repressive or restitutive (Lukes & Scott, 1984). Ethics is thus normative in the sense of being *constraining*. Thus, although ethical values are socially relative they are also societally transcendent, linking the individual members together through an often unrecognized but all-pervading evaluative framework.

In short, in becoming a "social fact" the ethical becomes inextricably linked to, and contained within, particular social settings. Ethics is socially relativized (normativized) and socially contained. But, more than this, as a social fact, ethics also functions itself as a pervasive and constraining (normative) force to support a particular status quo via sanction and praise. This second aspect of ethics' normative redescription, like the first, suggests a picture of society as an autonomous and bounded totality with a center and margins. The ethical plays a regulatory and homeostatic role in maintaining the equilibrium and bounds of the social organism, the importance of which is, of course, exemplified in Durkheim's famous study on suicide (Durkheim, 1977). According to Durkheim, those social settings that lack appropriate (moral) regulation are likely to induce feelings of anomie in susceptible individuals that can eventually lead to suicide.

It is difficult not to admire the theoretical symmetry of Durkheim's conception of ethics. His normative redescription of ethics is both a product of and a support for his methodological positivism and functionalism. It reinforces a conception of society as an autonomous and self-regulating entity and sociology as an autonomous and bounded discipline. It also apparently dissolves descriptive/evaluative tensions insofar as any ethic can only be evaluated in terms of its normative function in ensuring the continuity, and maintaining the bounds, of a given social formation. Thus, in Durkheim's view, the evaluative and the descriptive roles of the sociologist are fused. This is illustrated by his account of education, the function of which is "to arouse and develop in the child a certain number of physical, intellectual and moral states which are required of him both by the political society as a whole and by the special milieu for which he is specifically destined" (Durkheim in Lukes, 1988: 132). The state, as an embodiment of society, had a duty to inculcate suitable norms for individuals to follow and to provide appropriate educational institutions to ensure their propagation. Such norms were to be identified by sociologists who would also determine the best methods to ensure their educational transmission via schooling to the next generation.

The sociologist is then an engineer of society, a technician employing a technical rationality to ensure society's efficient functioning. This functionalism does not entail that society is entirely static. Durkheim's ingenious solution to the descriptive/evaluative dilemma places the sociologist in the position of a "prophetic expert," called upon to identify "the tendencies to movement, the germs of change, *aspirations* to a different form of society, *ideals* to be realized" (Durkheim in Lukes, 1988: 503). In recognizing and facilitating social trends sociology is not tied to a merely descriptive and conservative role as guardian, expositor, and reproducer of tradition but can also bring about moral progress and social evolution. Sociology is given the role of identifying current trends and articulating the changes necessary to ensure a smooth transition from one state to another. The sociological expert is given the right to divine and define the future.

In this way Durkheim's theoretical perspective incorporates and is dependent upon several of modernism's articles of faith; the ideas of gradual social progress, deferral to "objective" (social-scientific) expertise and a conception of society as an enclosed and autonomous totality governed by its own logic—a logic that sociology as a discipline is uniquely placed to reveal. Durkheim's functionalism can also, in its own way, be seen as an expression of the process of rationalization. The normative and functional redefinition of the moral field presaged in Durkheim's work effectively instrumentalizes ethics.

From Durkheim to Parsons: The Continuing Career of Ethical Normativism

As we have seen, the role of the sacred is both transformed and diminished in organic society. The ethical becomes flattened into a form of communicative cybernetics, playing a homeostatic role in connecting together parts of a society now regarded as a systemic whole rather than an undifferentiated mass.[11] This "moral cybernetics" appears most clearly in Durkheim's concept of "moral density."

At first, Durkheim's claim that moral density increases in organic society seems counterintuitive given the moral saturation of mechanical society in the conscience collective. This confusion is partly due to Durkheim's rather odd use of "density". To the physicist increasing density would suggest an increase in mass per unit volume, but Durkheim claims that in modern society both the social mass *and* the social volume increase. The volume increases as people make connections with distant regions, as economic and political systems interact; the social mass increases as the population grows and becomes more concentrated in towns and cities. Durkheim actually uses "moral density" as an indication of the number and complexity of social interactions, of what might in cybernetic terminology be termed the degree of social *feedback*.

Moral density has a dual role for Durkheim. First, it provides an explanation for bringing about the functional diversification that marks organic society: the denser society becomes the higher the levels of competition between individuals and the greater the pressure for diversification to avoid clashes. Second, it also has a regulative function: the higher the moral density the greater the degree of communication between this increasingly diverse population. In other words, as moral density increases, both the opportunity for, and the need of, increasing levels of communication and interaction become apparent.

In modern society the ethical has retained something of the pervasiveness that the conscience collective enjoyed in primitive society but has changed its mode of operation. It no longer regulates society by the ideological domination and subliminal exclusion of alternatives but instead functions to transmit information between parts of an increasingly complex social whole. This obviously helps to explain modernity's uncanny ability to avoid fragmentation in the face of the competing and partisan desires of individuals and sectional groups. However, the cost of retaining a normative role for ethics in organic society is to equate "ethical intensity" with "social density" in terms of intercommunicative activity. This has important ramifications. First, the ethical sphere loses its specificity: ethics becomes almost indistinguishable from any of the other communicative activities that serve to connect people. Second, what began as a sociologically satisfying account of the various manifestations and the authority of primitive religiosity ends as a restrictive and reductive framework that entirely subsumes the ethical under its normative (prescriptive) function. Ethics' inspirational vitality is smothered and the phenomenology of moral feeling ignored except insofar as it is regarded as a necessary epiphenomenon of social regulation. The heartfelt aspects of ethics—of love, care, and sympathy—are turned into a social currency to facilitate trade-offs between people with different interests and aims.

Durkheim stands at the head of a line of sociological functionalists, some of whom were to give an explicitly cybernetic twist to his theories—purposively declaring that society is, like biological organisms and complex machines, a system regulated by constant readjustments brought about by feedback mechanisms. (In so doing they once more exemplify the direct influence of material culture upon the form and content of theory.[12]) These feedback mechanisms ensure the survival of the whole by communicating information between parts acting as a form of cybernetic governor, a mechanism that ensures society's automatic self-adjustment to changing circumstances.

Foremost amongst these structural-functionalists was Talcott Parsons (1902–79) who had a profound impact on American sociology. Parsons saw society as composed of four hierarchically organized systems: the biological, personal, social, and cultural, each related to the other through the exchange of symbolic information. Ethics plays a vital role in integrating these systems; indeed, Parsons goes so far as to declare that "moral standards become the aspect of value-orientation which is of greatest concern to the sociologist" (Parsons, 1970: 14). Shared value-orientations ensure the smooth running of society by providing a recognised framework within which the interpersonal activities that make up the social system can operate. The values themselves are inculcated as the subject is "socialized" in the cultural system, that is, the highest and most general systemic level including aspects like language and national sentiment where "meaning is not contingent on highly particularized situations" (Parsons, 1970: 11).

While there may be some shared value-orientations Parsons' view is that at the personal level individual humans are out for their own self-gratification, they are very much the self-interested profit maximizers of economic and utilitarian theory. Parsons' later interest in sociobiology emphasizes that he believed that this attitude was based on *biological* factors. Interestingly then, Parsonian functionalism does not envisage an entirely closed social system as Durkheim did. Nature does play a role here in providing both the biological basis for the personality system and the environment against which society operates and, for this reason, some theorists have found some aspects of Parsons' cybernetic conception of nested systems amenable as a way of theorizing relations between nature and culture.[13] Unfortunately, for Parsons himself, nature remains something of a biological given and Parsons uncritically and unreflexively reads into nature the very assumptions that inform modernity's own shared value-orientations, for example, of naturally competitive individuals whose self-aggrandizement is thwarted only by the imposition of a form of cultural indoctrination. Nature, including human nature, is usually only referred to in order to prop up otherwise problematic assumptions about those nonbiological systems regarded as further up the organizational hierarchy.

Parsons also retains and develops Durkheim's progressive evolutionary model of societal development in a manner that is, if anything, more deterministic and less critical than Durkheim's own. He identifies a series of "evolutionary universals" that are, in order of appearance; social stratification, cultural legitimation, bureaucratic organization, money economy and markets, generalized universalistic norms, and democratic associations (Wallace & Wolf, 1995). These changing and developing patterns of social relationships are brought about, as in Durkheim, by the increasing differentiation of roles leading, via positive feedback, to a stratified society. While this social evolution is not always posited as a uniform and linear process there is no doubt that Parsons saw a liberal democracy as a logical endpoint for social development. In addition to ensuring the efficient socialization of individuals into society another of ethics' cybernetic roles is to ensure that changes within the system happen gradually and efficiently.

Parsons' reconfiguration of the moral field in terms of cybernetic governance leads, as many commentators have noticed, to a conservative and entirely functional account of society and ethics. Ethics' role is an entirely instrumental one, to ensure the efficient running of the social whole. This does not mean that Parsons was unaware of the many examples of moral rebellion within society, but he never seems to regard them as a challenge to the structures of society per se. Rather, they find themselves reinterpreted in terms of a form of feedback. They provide necessary information about social tensions that once expressed can be resolved, thereby enabling society as a whole to evolve further along an almost preordained route toward an increasingly rationalized social structure, that is, that of "modernity." To this extent protests against increasing rationalization, against the rule of efficiency and the normative dissolution of ethics, against the conceptions of (human and nonhuman) nature presupposed by Parsons are always going to be reinterpreted as self-defeating since they are by definition subservient to the interests of the current social order. From a Parsonian perspective there can only be change within and not of the social system.[14]

Habermas and Post-Durkheimian Ethical Theory

There can be no doubt that Durkheim's normative legacy in social theory makes radical environmentalism's task of producing a critical ethics exceedingly difficult. Indeed, Durkheim's normativism paves the way for a fully fledged instrumental and functionalist reduction of the ethical sphere. This conservative functionalism seems certain to follow in normativism's wake, especially when the sustaining power of the evolutionary metanarrative begins to lose credibility. Once "progress" is no longer deemed inevitable the emphasis shifts to morality's role in maintaining contemporary social stability and reproducing *this* society over forthcoming generations. Morality inevitably appears as a support for the status quo, constantly calling the individual to order and strengthening current social norms. Ethics is instrumentalized as a conservative field holding society together, mapping and patrolling society's borders and defining behavior as normal or deviant. Durkheim had of course himself always emphasized this prescriptive normativism claiming that "the external sign of morality . . . consists in diffuse repressive sanction, that is, blame on the part of public opinion" (Durkheim in Lukes, 1988: 410).[15]

However some theorists, most notably Jürgen Habermas, have tried to retain a cybernetic approach while acknowledging the existence of radical conflicts and disparate values in society. Although Habermas does retain something of modernity's myth of progressive societal evolution, he also recognizes that ethics is not necessarily normative in the prescriptive sense.[16]

For Habermas ethics is re-envisioned as a procedural discourse, an ethical talking shop, designed to redeem normative claims to validity in a world forced to recognize actors' competing interests and backgrounds (Habermas, 1990). Ideally this discourse should take place in an arena where uneven power relations between individuals are minimized and all are on an equal footing; what Habermas refers to as an "ideal speech situation."

Both Durkheim and Habermas regard ethical norms as tied to particular social situations, as societally relative, and both have a common conception of modernity as a space defined by and requiring an increasing moral density. Both allocate modern ethics a communicative and homeostatic role, ensuring the smooth running of a fragmented but increasingly interdependent society. However, Habermas takes Durkheim's normativism a stage further. Habermas wants to recognize the possibilities for ethical change and advancement introduced by the increasing reflexivity of individual and collective agents in modern societies. Where Durkheim believed morality to exercise its authority largely behind the backs of individuals, Habermas' discursive model presumes an ethical transparency and a degree of individual control that emphasizes that norms are constantly renegotiable. Habermas' variety of ethical normativism is less rigid than Durkheim's: while still regulative, it is cooperatively malleable rather than necessarily conservative. Such "discourse ethics" allows much more scope for critical perspectives and for moral change to be incorporated into society.

However, the specificity of the ethical relation is still lost in the cybernetic maze as the sacred becomes little more than a constantly revisable social compromise. As Habermas states, "the socially integrative and expressive functions that were at first fulfilled by ritual practice pass over [in modern society] to communicative action; the authority of the holy is gradually replaced by the authority of an achieved consensus" (Habermas, 1987: 77). What Habermas proposes is a "linguistification of the sacred" (Bernstein, 1995: 88–135) replacing the unspoken authority of society by the individual's obligations to the inbuilt rational requirements of participating in communicative discourse itself (McCarthy, 1984: 310–19; White, 1990: 48–58). Thus Habermas emphasizes the "performative contradiction" involved in engaging in a discourse without being willing to further justify one's statements when requested to do so. Where Durkheim envisages specific moral norms (e.g., individualism) as the outcome of, and solution to, the presence of competing interests in organic society, Habermas' discourse ethics recognizes the impossibility of universal ethical agreement at the level of specific norms and instead places its cybernetic faith in the transcendental linguistic principles behind its universalist pragmatics.

While making a moral principle out of the need to discursively justify all moral statements may have a certain intuitive validity in a democracy, and while Habermas' variety of normative ethics allows considerably more scope for moral critique, the ethical is yet further divorced from its non-normative manifestations. In effect, the individual becomes the "owner" of a set of values that, once translated into statements, become "tradable" within the ideal speech situation in order to reach a moral compromise. Yet, by definition, this is to reduce the sacred to the profane. It produces a system where morals are shunted about, transferred, and transformed from one site to another in a frenzy of continual and accelerating exchange. Indeed the ethical is, somewhat ironically considering Habermas' Marxism, reduced to the communicative equivalent of system of exchange values. (The myth of the ideal speech situation might equally be seen as the moral equivalent to that other modern myth, the "free" market.) Habermas presents us with a discursive equivalent of the economization of values. But the question remains as to whether ethical value is something that mere consensus has the power to impart, or whether such values are things we can simply trade (or talk) away.

Environmental Ethics and Functionalism

From a radical ecological perspective the ethical cannot be located entirely in the systemic interchanges between individual humans. Ethics also has to include our relations to nature; it is a lived multidimensional relation of care for natural (and human) others, a relation that originates in part from the environment itself. Yet nature has no voice in the ideal speech situation. In this way Habermas places himself firmly in Durkheim's sociological footsteps by maintaining that the ethical field is a cultural reality that is closed to nature.

Not surprisingly, Habermas is quite explicit in rejecting any ethical relation to the natural environment per se: "Human responsibilities for plants and for the preservation of whole species cannot be derived from duties of interaction and thus cannot be morally justified" (Habermas in Vogel, 1996: 159). His redefinition of ethics in terms of intersubjective communication between human actors maintains the social and ethical closure instigated by Durkheim's social model.[17] Although the rationale for this ontological and epistemological closure is redrawn in terms of a procedural humanism rather than a methodological positivism, the effect for nature, and the radical environmentalism that would speak of it, remains the same.

Part of the problem for environmentalism is that it tries to say to society those things that society finds most difficult to hear. If we take seriously the points I have been making about the intimate relation between social formations and theoretical (discursive) forms, then we must recognize that there are aspects of an environmental ethos that may even be ineffable in our present language and society. Habermas underestimates the normative and regulatory function of language itself as a structure that facilitates the transmission and reception of certain messages but also necessarily restricts what it is possible to say. Not only can language be ideologically conservative but by reducing the mode of transmission and exchange of ethics to (human) language Habermas further constrains the moral field within a set of explicitly anthropocentric boundaries (Whitebook, 1979).[18] Ethics becomes a wholly cultural phenomenon and one that can only be acknowledged insofar as it can be voiced. But nature communicates to us, and through us, in many different ways; through the early morning's light, the night's warm breeze, the bark's rough texture, the dawn chorus, and so on.[19]

In short, while all of these normative and cybernetic approaches, from Durkheim to Habermas might help grasp something of the ethical dynamics of modern societies, they do so only at the cost of furthering the rationalization and instrumentalization of the moral field. Morality is constricted and distorted as it becomes subsumed under prescriptive frameworks; it loses much of its oppositional and phenomenal multidimensionality and becomes an abstract managerial tool for ensuring social homogeneity rather than a source of "sacred" or heartfelt opposition to the destructive trajectory of modernity.[20]

What is more, Durkheim's, Parsons', and Habermas' redrawing of the moral field in normative and functionalist terms depends to a great extent on those shared assumptions identified earlier, that is, ideas of social progress, social-scientific expertise, and societal and sociological autonomy. But radical environmentalism challenges all three of these assumptions.[21]

In terms of "progress," environmentalism, like many other radical movements, tells a different "history" (Merchant, 1990; Seymour & Giradet, 1990; Ponting, 1991; Sale, 1991; Grove, 1996). Rather than extolling the virtues of that continual linear advancement supposedly brought about by technological mastery and human ingenuity, it recalls the cost in human and environmental terms of each "advance." Its anamnesic critique reminds us of that cultural and environmental holocaust which has dogged modernity's passage and the fate of those communities, from Mesopotamia to the Maya and from Easter Island to Three Mile Island, that fail to give due regard to their environs.[22]

Environmentalism also questions the right of "experts," whether natural or social scientists, to determine our collective futures (Shive, 1994; Irwin, 1995). Bhopal and Chernobyl, BSE and botulism all exhibit the fallibility of such expertise and have undermined the lay public's trust in received knowledge. Indeed, awareness of such fallibility has become an integral part of our ability to reflexively recognize the dangers of our current existence, a situation characterized by Beck as living in a "risk society" (Beck, 1992; Giddens, 1990).

Perhaps most importantly, environmentalism questions the idea of modernity as a coherent and self-creating whole. Of course, critical theorists of many persuasions highlight modernity's internal fractures along lines of gender, race, creed, wealth, nationality, language, symbolic capital, and so on. They portray a society riven by inequalities, conflicting interests and power imbalances; a society that, far from exemplifying organic interdependence, relies on exploitation and competition, where parasitic rather than symbiotic relations prevail. From these radical perspectives modernity is falling apart at the (ethical) seams. But environmentalists add yet another, and in some respects a more radical, destabilizing critique—subverting modernity's social coherence from the outside—emphasizing society's dependence upon the "external" environment, the alien "other" of nature.[23]

Normative Ethics and Sociology's Disciplinary Boundaries

Before proceeding to try and recuperate something from these functionalist and neofunctionalist approaches it is worth bringing together some of the issues covered in order to identify the manner in which sociology, like philosophy, has come to incorporate aspects of the rationalization process (including the very need for disciplinary differentiation).[24] I have argued that, in many respects, Durkheim was the key figure behind the founding of sociology as a discipline, the subsequent disciplinary redefinition of ethics in normative terms and the closure of the social to nature. While sociology's disciplinary marginalization of nature is perhaps easily understandable, the rationale behind sociology's almost universal acceptance of a normative account of ethics is less immediately so.

The lack of alternative theoretical approaches to ethics is somewhat surprising since ethical issues and disputes pervade issues like social deprivation, social deviance, and social development. While ethical values and ideals critical of the status quo quickly assume central roles in everyday discussions of such issues, they seem to silently vanish as sociology approaches, leaving only tenuous impressions between and behind the lines of sociological texts.[25] While explanations and social typologies couched in terms of power and poverty, class or conformity, self or state abound, while gender, economic, and ideological relations are made explicit, morality remains generally shrouded in atheoretical mists. As Steven Lukes says, "[i]t is astonishing to see how little attention has been given to [moral] questions in twentieth century sociology and social anthropology. Indeed it is not an exaggeration to say that the sociology of morality is the great void in contemporary social science" (Lukes, 1988: 432).

Ironically it was Durkheim's sociological problematic, which both placed ethics center-stage and initiated that very process that resulted in the subsequent disciplinary marginalization of critical ethical frameworks. This marginalization follows from both his influence in defining sociology's disciplinary boundaries and instigating those normative trends that would later come to dominate sociological perspectives.

As Giddens notes, Durkheim began his career with the intention of founding a "science of morality" (Giddens, 1986: 63). This project became subsumed within his attempt to found a social science and in this sense the bounds of sociology also came to demarcate the limits of morality. As a discipline sociology has to describe and explain the manner in which particular societies are organized and operate. Among the things it must account for are the moral values of those societies. However, insofar as it wishes to account for these values objectively, that is, insofar as it wishes to become a social science, it must refuse to take sides over particular issues of right and wrong. That is, it must distinguish its meta-ethical theorization of ethics, what ethics is and how it functions, from particular ethical norms, what any given society deems right or wrong.

Philosophy of course makes a similar distinction between normative and meta-ethical theories but the need for this distinction is both more acute and less easily defended for a social science.[26] As the first chapter showed, philosophers are generally quite happy to use particular theories of what values are (meta-ethics) to criticize the actual values held by those who they consider to be (normatively) wrong. For example, Singer takes a meta-ethical position that argues that moral considerability is dependent upon one's sentience. He then uses this meta-ethical position to criticize those whose moral norms don't take certain animals into account. But, insofar as they wish to retain the mantle of objectivity, the social scientist does not have the luxury of simply declaring certain normative values to be "mistaken." The purpose of their meta-ethical speculations is not to engage with any particular normative perspective but to account for *all* values present in a given set of social circumstances. For example, where a philosopher might dismiss a certain set of religious ideals on the ground that the existence of God is ontologically implausible, the social scientist accepts that the social totality sets the limits for theoretical speculation and critique. In accepting disciplinary boundaries social theory's concerns become entirely delimited by social ontology, that is, what is present in that society. In this sense questions about the actual existence of God would be immaterial to sociological understandings of religious ideals and their questioning on these grounds is drawn out of (sociology's) bounds. No matter what their philosophical status, as "social facts" religious ideals remain a valid topic of sociological study.

Thus, for reasons of disciplinary closure and in order to retain an "objective" aura social theory happily endorses a theory of normativity that redefines ethics in line with sociology's interests, that is, in terms of morality's functioning within the social totality. This rationalization of the moral field takes a different form from the philosophical axiologies discussed in the first chapter, but in its own way it too embodies a formal (instrumental) rationality. Ethics is redefined in explicitly instrumental terms, as a means to society's ends: every moral act becomes a means of increasing social efficiency rather than being understood in terms of any possible intrinsic value. Indeed the very possibility of intrinsic values is excluded since all moral activities are understood only in terms of their extrinsic relation to society's functioning. In this way, despite claiming to be neutral and to make universally valid claims about the nature of morality, this cybernetic meta-ethics imposes its own (formal) rationality on the moral field. This supposedly neutral and objective perspective is nothing of the kind. While claiming to be neutral in terms of ethical norms, it actually replaces and excludes alternative meta-ethical understandings, understandings that provide the bases for different normative values. To return to our example, sociology cannot claim to stand back and simply describe religious ideals "objectively" since there is a very real conflict between the meta-ethics that ground religious ideals and those that ground sociological understanding. The meta-ethical rationale behind functionalist and positivistic sociology is that of instrumentality, of formal rationality itself. The meta-ethical redescription of ethics made by social theory in terms of their normative (i.e., homeostatic) function simply masks sociology's role in establishing a competing, rather than a neutral redefinition of the moral field.

In this sense then, some of the causes for ethics' normative reconfiguration lie partly within sociology itself, in the manner in which it has demarcated its area of study, fought its disciplinary corner, and debated its purposes and ambitions. But the root cause behind sociology taking this direction is not internal. Rather, it lies in the particular way that sociology, like philosophy, comes to incorporate and express that same process of rationalization that, as chapter 1 illustrated, characterizes modernity. These close ties with formal rationality are *implicit* in the very presuppositions behind developing a (supposedly) neutral social-theoretical understanding of ethics and *explicit* in the sense that the rationale behind having a discipline of sociology in the first place is justified "instrumentally" by its making society function more efficiently. All of the cybernetic approaches from Durkheim to Habermas have, in one way or another, this "technical" end in mind and to this extent their problematics embody a formal rationality.

We must therefore recognize that theoretical problematics of sociology, like those of philosophy, are embedded in the society they seek to understand and that certain strands in particular are wedded to an instrumental rationality. At this point though we should recognize that it would be a mistake to regard either philosophy or sociology monolithically. There are different degrees of involvement in modernity, different strategies, and different interests at stake.[27] Certain strands of social theory reject both cybernetic approaches and the idea of an objective social science. These strands are, like radical environmentalism, engaged in a fundamental critique of the current social order and of the presuppositions underlying the ontological closure of society and epistemological (disciplinary) closure of sociology. An understanding of these critiques is absolutely necessary if environmentalism is to prove capable of opening a multidimensional and permeable ethical space for those nonhuman others who have fallen by modernity's wayside.

Social Morphology and an Ethics of Place

Insofar as normative perspectives tend to underemphasize the permeability of social boundaries to environmental influences they fail to make the conceptual links necessary for articulating an environmental ethos. However, just as the road's surface is fractured by those seemingly delicate "weeds" that push up through the asphalt's layers, so a theoretically repressed nature has a tendency to return in unexpected ways and shatter theory's inattentive complacency. I want to finish this chapter by returning to Durkheim for, despite his best efforts, even Durkheim failed to make his social theory water-tight against nature's intrusions. Thus certain aspects of Durkheim's work, especially his idea of "social morphology" might still help revise conceptions of society and ethics as enclosed totalities.

The reasons for this are twofold. First, social morphology has a somewhat ambiguous and perhaps even subversive role in Durkheim's sociological problematic and for good reason. It represents Durkheim's only real recognition of both society's dependence upon external "geographical" influences and sociology's relations to other disciplines. It undermines the absolute autonomy of society and sociology. Second, it might be regarded as an incipient spatial metaphorics, as a way of beginning to spatialize our social, ethical, and environmental relations. This fits with Lefebvre's remark that "space *is* social morphology" (Lefebvre, 1994: 94). It might also fit with Giddens' suggestion that "we should reformulate the question of order [i.e., normativity] as a problem of how . . . social systems 'bind' time and space" (Giddens, 1990: 14). While I shall leave the development of a spatial metaphorics for later chapters, I want to illustrate here once more how taking nature into theoretical account will alter the form our problematics take.

Social morphology is defined rather vaguely by Durkheim's nephew and successor Marcel Mauss as "the mass of individuals who compose the society, the way in which they occupy the land, and the nature and configuration of objects of every sort which affect collective relations" (Mauss & Beuchat, 1979: 3).[28] Social morphology is a kind of substratum for social life. However, it always occupied a marginal place in Durkheim's sociology; indeed it gave "the appearance of having been created as an afterthought, put together from various subjects" (Fox in Mauss & Beuchat, 1979: 5).[29] This vagueness and eclecticism is, I would suggest, inevitable given social morphology's role of articulating an epistemologically closed sociology and an ontologically enclosed society with nonsociological and asocial factors like geography and the natural environment. Social morphology, Mauss hoped, might allow sociologists to both have their disciplinary cake and eat it. While it remained "a part of sociology, virtually its primordial half . . . it is also one of its most independent parts:" While a distinct domain *within* the social domain "it also unites various sciences that are ordinarily but unduly separated" (Fox in Mauss & Beuchat, 1979: 5).

It was this geographical (or environmental) relevance that led Mauss to develop the idea of social morphology at length. Though it lost little of its vagueness, and while Mauss continued to distinguish his own sociological work from that of his geographical contemporaries, he recognized that sociology could not pretend to exist in isolation and environmental factors could not be ignored. Indeed, Mauss saw his work as a necessary sociological corrective to competing anthropogeographical works that, he claimed, tended to reductively describe and explain societies in terms of factors such as land. As Mary Douglas recognizes, Mauss used social morphology as part of "an explicit attack on geographical and technological determinism" and as a demand for "an ecological approach in which the structure of ideas and of society . . . are interpreted as a single interacting whole in which no one element can be said to determine the others" (Douglas in Mauss & Beuchat, 1979: 11).

Mauss believed that his study on the seasonal variations of Inuit society revealed distinct social morphologies during winter and summer. In the summer, he claimed, Inuit society was widely dispersed, each family unit living together in tents. In the winter the families become conscious of themselves as a group through the intense social activity that occurs while they congregate together in concentrated settlements and communal buildings. These changes in social morphology also affected ethical morphologies, for instance, the winter being marked by communal rather than individualistic responsibilities and sexual practices. Of course the seasonal change was not complete; nor did Mauss suggest that these changes were entirely due to environmental influences, they were socially reproduced. Nonetheless, the differences were marked and affected all aspects of that society whether economic, political, symbolic, or ethical.

I want to suggest that Mauss' development of Durkheim's terminology enables us to see how terms like "social morphology" might operate to allow us to posit a dialectical and genuinely multidimensional relationship between a nature and culture that are only ever relatively autonomous. We need not abandon entirely Durkheim's insights into the supra-individual origins of ethics but, once we free Durkheim's concepts from a role in policing his positivistic requirements for complete ontological and epistemological autonomy, we open a way toward an environmental ethos. This means going further than Mauss did in recognizing the permeability and interdependence of nature and culture and in acknowledging the often competing forces that shape the moral spaces we inhabit. It also requires a radical revision of Durkheim's normative understanding of ethics.

As we have seen, ethics for Durkheim is normative in both the sense of being socially relative and socially regulating, it is *external*, *general*, *and constraining*. However, I have suggested that ethics is not only *external*, but has an "internal" phenomenology, values are felt and nurtured within as well as without. The individual does not passively imbibe social norms along with their mother's milk. Morality entails an active engagement in the world, it is a relational activity requiring that the individual recognize tensions and on occasion take sides. From this it also follows that ethical fields are not necessarily *general*, but may be peculiar to certain situations and localities. Moral values and understandings are not dispersed evenly throughout society, they are not universal and all pervasive, but are nodal and rhizomatic.[30] They are, it is true, always more than the product of individual whim, but they grow and develop relationally in concert with particular settings. Nor is ethics necessarily *constraining*—it need not entail punitive measures or the imposition of sanctions on others. Indeed, I will suggest, an ethical relation is primarily one that recognizes in others their requirement for *a space of their own*, for the *room* to develop according to their own aspirations and needs. An ethics of place in particular takes responsibility for the creative and careful protection or reconstitution of appropriate spaces for natural (and human) others.

It is ironic that, while Durkheim's normative approach correctly recognized the need to see ethics in relation to particular social settings, like many of his social theoretical descendants, he lacked a reflexive understanding of the origins of his own, supposedly objective, ethical problematic. He failed to see how his normative redescription of the ethical sphere could itself be interpreted as a product of a society obsessed with regulating its members and an academic sociology obsessed with disciplinarity. Similarly the later reformulation of ethics as cybernetics also mirrors technological and structural changes in modernity (Robbins, 1995).

If a normative (i.e., prescriptive) conception of ethics is a product of our contemporary social formation, then it is unsurprising that those seeking to make radical social changes will find it fails to reflect their understanding or concerns. A radical environmental ethic would not seek to regulate, to impose duties, to constrict the space of the other, but to let things be and become, to create spaces and protect places that give the autopoeitic activities of nature an opening to unfold. It would not wish to center values and behavior in current structures but would emphasize decentralization and attentiveness to locality. The abstract disembedded moral frameworks characteristic of modernity posit an "empty space" (Giddens, 1996: 18)—they separate space from place in order to facilitate ethical decision-making at a distance. An environmental ethic must reconnect and re-embed ethics in natural *and* cultural places. Rather than operating to reduce the difference between things, an ecological understanding recognizes the import of *différance* in the Derridean sense—that relational spacing of things that gives them meaning by recognizing their position in respect of (and for) each other. An environmental ethic celebrates the multidimensionality of our relations with nature and resists attempts to reduce them to abstract rubrics or a cramped cybernetics. It promotes an an-archic attentiveness that is the very opposite of rule-bound regulatory frameworks of current ethical perspectives. Above all, it rejects the view of society as a self-contained sphere and seeks to widen those small fissures in its concrete coat, to make permeable and dissolve its absolute boundaries so we might touch and be touched by nature.

3

Social Theory, Nature, and the Production Paradigm

The last two chapters have suggested that theory does not just describe but actively reconfigures the ethical field, shaping it to fit its own concerns. Those theoretical perspectives that claim to capture and elucidate certain essential features of morality within neutral sociotheoretical or philosophical frameworks are mistaken. Their abstract theoretical forms are not the mark of their situational transcendence but of their embodiment and expression of that formal rationality that characterizes modernity. They are, in Marcuse's terms "positive" philosophies insofar as they comply with, support, and express the normative values of the current social order.[1]

Having previously outlined some of the problems with contemporary ethical theories this chapter will turn to the need to reimagine the place of "nature" as the other component of a radical environmental ethos. The theoretical difficulty here is, I shall claim, exactly analogous to that facing ethics, namely that the predominant attempts to conceptualize nature accord with those modern worldviews that have proved so damaging to it. In particular, nature is regarded as something external to society and as a passive resource awaiting exploitation, that is, it becomes subject to an instrumental rationality.

The problem is how to escape from these instrumental theoretical frameworks and reimagine nature in other ways. As critical theorists like Marcuse recognized, all theory is, to some extent, a prisoner of its circumstances. One cannot think wholly outside of what is "given" and, even if it were possible to escape the confines of our current language, ideology, economy, and so on, the need to communicate intelligibly with others would still require that thinking be grounded, to some degree, in present realities. Nonetheless a critical theory has to attempt to go beyond what is simply present, to transcend the given, and thereby confront society with possibilities that reside elsewhere. In this way a critical theory can challenge the one-dimensionality of contemporary thought and practice.

From this critical perspective the theorist has to recognize that, like it or not, her thought is politically and ethically situated in relation to the current social order. Every utterance takes on social import in terms of its either furthering the cause of the dominant ideology or striving to fracture its uniformity. This is what Louis Althusser meant when he defined philosophy as politics in the field of theory (Althusser, 1993: 108).[2] A philosophy can either slip easily into the interstices of the social machinery, oiling its cogs and ensuring its smooth running, or it can act as a spanner in its ideological works making us suddenly aware of our overdependence on mechanisms we normally take for granted. It is this theoretical "monkey-wrenching" that might aid radical environmentalism in its opposition to our destructive social order and those equally destructive patterns of thought that serve to support it.[3]

However, society is not mere machinery, its complex workings are infinitely more difficult to fathom and the success or failure of any theoretical intervention depends on its ability to both challenge and strike a chord with current thinking. A new problematic must both adopt and adapt, appropriate and expropriate, concepts from its forerunners and competitors, thereby changing their theoretical specificity. And here perhaps I need to say a little more about the term "problematic." This too is borrowed from the work of Louis Althusser who claims that any theoretical discourse will consist of a framework of concepts in structural relations to each other—relations that dictate both the meaning of those concepts, and the questions, interests, and presuppositions appropriate to the theory as a whole.[4] These relational frameworks, or problematics, exclude certain possibilities from consideration, make some concepts central, others peripheral and so create a position, "a particular unity of a theoretical formation," that "binds" those who use it. The problematic is not just a theoretical tool to be applied to the world but the framework in which problems develop and within which proposed solutions are judged. A problematic both creates and emphasizes particular theoretical relations, it both facilitates and constrains the way in which we understand the world.[5] The concept of a problematic thus supports a view of theory as an active agent refashioning the conceptual field from which it emerges and with which it engages. (Though of course theory is only ever one aspect of this refashioning, which also depends upon practical, institutional, and traditional interventions and interests.)

Having engaged in a critique of ethics, as conceptualized in certain modern philosophical and social theoretical problematics, I tentatively suggested that a spatiotemporal reading of ethics might assist environmentalism in its dual tasks of fracturing the current conceptual consensus and reimagining a specifically environmental ethics. Without pre-empting the work of later chapters, the problematic I seek to develop will speak of the production of ethical spaces and in particular the role of the natural environment in this productive process.[6] In other words I want to adopt and adapt a productivist framework as a basis for a critical theory of ethics and the environment. But this strategy is problematic in a different sense since the emphasis on nature's instrumental values is nowhere more prominent than in the productivist paradigms that I hope to employ. Examining the extent to which productivism itself might be tied to modernity and modern attitudes toward nature is therefore crucial if my own productivist metaphorics is not to fall prey to the criticisms I have leveled against others.

In particular I want to examine this issue in the context of Marxism. There are a number of reasons for choosing to investigate Marxism's theoretical relationship with "nature." First, because Marxism utilizes an explicitly productivist theoretical framework. Second, because as a major strand of social theory that situates itself in terms of a critique of current social and economic relations, and being closely tied to critical theory, Marxism might be thought to have much to offer radical environmentalism. Third because of a recent surge in interest in "the environment" and environmentalism by Marxist theorists there is a wealth of material available exhibiting varying degrees of sympathy/hostility toward environmentalism and varying degrees of flexibility in their attempts to account for nature within the bounds of productivism.

It is clearly beyond the scope of this chapter to recount the entire history of the conceptualization of nature in Marxist thought or to attempt to systematically cover all the many interventions of Marxists in environmental debates (but see Schmidt, 1971; Leff, 1995). Instead, I intend to provide a conceptual outline of Marxist productivism and show how the acceptance of this particular problematic configures and delimits the idea of nature. In this manner it should also become obvious where the lines are being drawn in debates between Marxism and radical environmentalism and what repercussions these differences have for our idea of nature.

Marxism and the Productivist Paradigm

The concept of production has retained a central role in all varieties of Marxism. Traditionally, economic practices mediated by industrial or agricultural labor are the paradigmatic forms of the productive process.

In the recent past, the specificity of this economic paradigm lent Marxism an air of authority, underlining its relevance to the plight of the working classes in modern capitalist societies. It was easy to connect Marx's theoretical account of capitalism with the everyday experiences that constantly brought home the injustice of a system where those whose labor created the material basis of society found that the labor process also effectively estranged them from that society. Capitalism ensures that we each become involuntarily tied to a particular sphere of productive activity that is far from fulfilling either our needs or expectations. In Marx's words "this consolidation of what we ourselves produce into an objective power above us, growing out of our control, thwarting our expectations, bringing to naught our calculations, is one of the chief factors in historical development up till now" (Marx in Ricouer, 1986: 85).

For Marx, productive activities provided the key to understanding a society's operation. But making labor *the* locus of productive activity can and often does lead to a reductive economism. Although, in one sense, Marx is absolutely right when he states that "[m]aterial production . . . is the basis of all social life" (Marx, 1990: 286), if the emphasis is placed entirely on economically productive labor (rather than say certain cultural activities) the economy inevitably comes to be regarded in terms of the base upon which the entire social superstructure is raised. Thus, we can easily fall into a variety of determinism where the economy is seen as the only important controlling influence in society.

The debates surrounding such economism are familiar enough to not require further elaboration here (Williams, 1989). Treating ideology, culture, politics, and the like as mere epiphenomena tacked onto the economy, as economic by-products with little or no influence or import, is obviously problematic. It is doubtful that Marx himself ever held such a simplistic unidirectional causal model despite occasional statements to the effect that the "hand mill will give you a society with a feudal lord, the steam mill a society with the industrial capitalist" (Marx, 1977: 389).[7] Marx certainly does recognize, and go into considerable detail about, the influence of social circumstances (the superstructure) upon the production process. But whether or not Marx was himself advocating a variety of economic determinism, there is little doubt that the metaphor of production that underlies his entire problematic accumulated much of its political capital precisely because of this close relation to economic production and industrial labor.

One of the reasons why an economistic version of productivism has increasingly been subject to intense critique is precisely because, as industrial labor seems to play an increasingly peripheral role in developed Western societies, this very specificity has become more and more detrimental to the development of Marxist theory and practice. In a society where the division of labor has continued apace, where new technologies have arisen, where the cultural field seems ever more influential, many critics have felt that an economically reductive framework no longer has the explanatory power necessary to grasp the complexities of contemporary society.[8] The corollary of this is that Marxism no longer has the intuitive appeal that it once had for a working class that often no longer recognizes itself as such and is perhaps less willing to define itself primarily in terms of economic roles. The emphasis on the economy previously regarded as inviolable by traditional Marxists has increasingly come to be regarded by critics on the left as well as the right as an anachronistic shackle on thought.

Given an increasingly skeptical populace, and since productivism is at the very heart of the Marxist problematic, traditional Marxism faces something of a dilemma. Should it continue to stick with a more or less economistic model and argue for its continued relevance or should it emphasize the existence of other aspects of production? The past few decades have seen this debate between theoretical orthodoxy and innovation carried out on a number of contested terrains, in particular between Marxists and other "nonclass"-based political movements such as feminism. Thus, for example, in response to feminists claims that Marxism had undervalued those activities of women such as child-rearing and housework that are accomplished largely outside of mainstream economic processes, Marxist apologists tried to link women's contemporary social roles to the capitalist mode of production (e.g., Vogel, 1983). But, insofar as Marxism wishes to prioritize economically determined features like "class" while feminists call for the recognition of patterns of domination based on gender their respective problematics are always likely to be incompatible to some degree. Even the most optimistic assessments of this attempt to wed Marxism and feminism regarded its results as something of an "unhappy marriage" (Sargent, 1981; Landry & Maclean, 1993).

Certain Marxist theorists tried other innovative ways of getting around this problem. Althusser famously tried to retain the economic specificity of Marxist analysis while making much of the relative autonomy of politics, ideology, and theory. He argued that each one of these "levels" of society is distinguished by "the type of object (raw material) which it transforms, its type of means of production and the type of object it produces" (Althusser & Balibar, 1986: 59). Thus, for example, theoretical practices (sciences) work on the inadequate abstractions (raw material) that arise in our day-to-day involvement in other nontheoretical practices to produce knowledge.[9] The connection between theoretical and economic realms is not severed, since some of our inadequate abstractions arise in our practical experiences of economic production, but it is made much more tenuous. This is also true for the political and ideological levels and Althusser claims that each level is only economically determined "in the last instance." And, from "the first moment to the last, the lonely hour of the 'last instant' never comes" (Althusser, 1969: 113) since the economy is never found in isolation from the superstructural elements.

The advantages of Althusser's reading of Marx are obvious insofar as he nominally retains the priority of the economic sphere, and of Marxist social analysis, but actually allows much greater leeway for identifying other factors that might exert an influence within the social superstructure. However, few were happy with his compromise, seeing it as either an overtheorized challenge to traditional Marxist politics or, from the opposite extreme, as a mealy mouthed attempt to salvage an orthodox Marxism from its immanent demise.[10] For the purposes of the present chapter, Althusser is interesting insofar as he enlarges and alters the boundaries that previously defined the concept of production. In effect production becomes applied to a much wider range of activities than the standard economist reading would allow. In granting a relative autonomy to different social practices, economically productive work ceases to be the sole archetype of labor and the range of materials utilized and produced without the direct intervention of the economy is vastly increased. Althusser's approach also leads to an emphasis on the material reality of both ideology and theory, that is, of ideas, values, theories, and so forth.[11]

Althusser's attempt to broaden the concept of production seemingly gains support from comments made by Paul Ricoeur. Ricoeur argues that there are passages in Marx's early works, such as the *Economic and Philosophical Manuscripts*, which would seem to imply a rejection of any reductive economism. Ricoeur suggests that, when Marx states "that religion, family, state, law, science, art, etc., are only particular modes of production and fall under its general law" (Marx in Ricouer, 1986: 59) he does not mean that the economic base determines these superstructural elements. This interpretation would be to read economism back into the text where none originally existed. Instead, Ricoeur interprets this statement as developing an analogy between economic and other practices in order to grasp their basic similarities. In other words, law, science, and art might all in their own ways be seen as productive processes, as creating their own "products."

This productive analogy obviously carries with it overtones of the nineteenth-century factory system but, Ricoeur claims, the term *producktion*, used by the early Marx, has a much wider application than simply the physical transformation of an object by labor. "In German the word *producktion* has the same amplitude as objectification—thus Marx's statement does not express an economism," though Ricoeur admits, "[t]he reductionism of classical Marxism is nevertheless nourished by the word's ambiguity" (Ricouer, 1986: 59). In other words, Ricoeur believes that the later Marx and many of his followers came to take the economic overtones of his productive analogy too seriously. The political capital to be gained by this maneuver was obviously an influential factor and in attempting to disassociate himself from his Hegelian and Feuerbachian influences Marx shifted his problematic inexorably toward an economization of society. Marx seemingly becomes progressively more dependent throughout his works upon a narrower and more economistic use of "production." "Production," once reified in this manner, lost its more general implications and instead its economic signification became fossilized in the base/superstructure model.

The implications of Ricoeur's claim are far-reaching. The concepts "production" and "objectification" have much in common, both referring to the processes by which objects are brought into being for a society and made social realities. But here, Ricouer suggests, the similarity ends, for objectification is not a process limited to the economic sphere but one that occurs continuously in our cultural life too, in our ideological, political, artistic, and theoretical endeavors. Each and every form of activity brings into being or reveals its own kinds of objects. This broader conception of production is exactly what Althusser requires to maintain the relative autonomy of his social levels. But there is an irony here since the analogical conception that Ricoeur finds in Marx's early philosophy seems unavailable to Althusser precisely because of his wish to maintain the break between the pure theory of the later Marx and the ideology of his earlier works. Althusser always regarded his project as one of reclaiming the more mature and scientific work of Marx's later years, from the idealistic influence of Hegel on his earlier work.[12]

Despite Althusser's refusal to use the conceptual repertoire present in the theories of the early Marx, his understanding production as occurring on different levels does gel neatly with Ricouer's point about understanding production as being equivalent to objectification.[13] Both perspectives seem to provide ways of avoiding the pitfalls of economic reductivism. The importance of these debates about the definition and remit of the concept of "production" will become more apparent as I turn to examine the relationship between productivism and nature.

Productivism, Labor, and Nature

There are, I think, two inseparable issues here that merit further investigation in terms of their implications for nature. First the question we have already begun to address, of how widely or narrowly one interprets the concept of "production," and second the question of how widely or narrowly one interprets the concept of "labor."

For Marx, labor is "an exclusively human characteristic" (Marx, 1990: 284), an activity, which demarcates us from the rest of nature; to labor, is to be human. He carefully distinguishes human labor and the work which other animals might engage in. "A spider conducts operations which resemble those of a weaver, and a bee would put many a human architect to shame by the construction of its honeycomb cells. But what distinguishes the worst architect from the best of bees is that the architect builds the cell in his mind before he constructs it in wax" (Marx, 1990: 284). Now this anthropocentric distinction is by no means supported by our knowledge of animal behavior: the ability to develop some sort of mental "plan of action" (rather than, say, just responding to stimuli via preprogrammed instinctual responses) is indubitably shared by other mammals like apes and dolphins.[14] (Darwin's ingenious experiments with leaf shapes proved to his own satisfaction that even earthworms had a rudimentary capability for abstract thought [Darwin, 1881]).

But Marx's definition doesn't just epitomize a prevalent anthropocentrism, it also tells us something about the notions of production and labor that he utilizes. His acceptance of labor as the key human activity dictates the economistic limits of his productivist problematic. Yet this does not mean that Marx is a straightforward economic reductionist. His definition of production and labor are not inflexible and, despite Ricoeur's analysis, the Marxian concept of production still retains something of the idea of objectification. The theme of objectification recurs throughout even his most mature works. For example, Marx claims that, in the process of changing the form of those materials they work upon, people realize [*verwirklicht*] their purposes in those materials (Marx, 1990: 284). And again: "A use-value, or useful article, therefore, has value only because abstract human labor is objectified [*vergegenstandlicht*] or materialized within it" (Marx, 1990: 129).[15] Each individual's thoughts become materially present in the new object that their labor has produced. This is why the labor process is also a process of self-realization. While it is true that objectification occurs only in the economic sphere, it is not, as Ricoeur suggests, a question of Marx defining production in terms of either objectification or the economy but of defining it such that objectification and the economy are both one and the same process mediated through purposive human activity, that is, labor. In this way each of Marx's concepts supports the other in a theoretical structure (a problematic) that depends upon the anthropocentrism inherent in the concept of labor to ensure its unity. Labor, as purposive and transformative activity, mediates between the subjective world of the human mind and the objective realm of nonhuman nature. The ends of this mediating process are dictated by human consciousness and the means by which it is carried forward is human labor.

This problematic limits what can and can't be said about nature and it is because of this emphasis on the dynamics of *human* labor that nature finds itself marginalized in spite of Marx's materialism. Nature may be present in those goods produced by labor but because these products are marked by the labor that produced them they are no longer natural. Though human labor too is in one sense a natural process—the act of demarcating human from nonhuman labor sets it, and the things it produces apart. The products of human labor are quite literally artificial—the results of human design and activity. (This designation as artifacts is after all what sets the bee's cell apart from the architect's labors.) This necessarily means that for the Marxist only the artificial ever has any value, nature in itself is always valueless.

Marx's productivist paradigm thus has profound implications for environmentalism. It seemingly concentrates all values within social processes—and in its narrowly productivist form—within economic processes. Nature only enters the Marxist equation through its participation as part of the "means of production," either as the "object of labor"—the raw material, or as some part of the "means of labor"—the tools utilized by labor to work upon that raw material, such as stones for grinding, pressing, and cutting. In neither case does it enter untransformed for, strictly speaking, nature as itself does not even attain the status of raw material. (This term is reserved for material that has already been worked upon in order to become an "object of labor," e.g., iron ore, lumber, etc.) Even "animals and plants which we are accustomed to consider as products of nature, may be, in their present form, . . . the result of a gradual transformation continued through many generations under human control and under the agency of human labour" (Marx, 1990: 288)[16] they are therefore no longer natural.

Of course, this emphasis on human intervention does not mean that labor can shape nature just as it wishes. If materialism is to mean anything at all then it has to recognize, as Alfred Schmidt suggests, that "[t]he stuff of nature, which Marx equated with matter, is in itself already formed" and that "[m]an's [*sic*] aims can be realized by the use of natural processes, not despite the laws of nature but precisely because the materials of nature have their own laws" (Schmidt, 1971: 63). Humanity has to work with and within the structural reality of the natural world. But although these "laws" both constrain and facilitate social activities, they are, as it were, turned against nature itself in the course of the labor process. The purpose of labor is to subdue nature, to make it serve human ends. "Labour is, first of all, a process between man [*sic*] and nature, a process by which man, through his own actions, mediates, regulates and controls the metabolism between himself and nature. He confronts the materials of nature as a force of nature. He sets in motion the natural forces which belong to his own body, his arms, his legs, his hands, in order to appropriate the materials of nature in a form adapted to his own needs. Through this movement he acts upon external nature and changes it, and in this way he simultaneously changes his own nature. He develops the potentialities slumbering within nature, and subjects the play of its forces to his own sovereign power" (Marx, 1990: 284).

Marx's language here is uncompromising. The consciousness that lies behind all labor, and distinguishes it as a uniquely human activity, purposively plans and executes a campaign to subvert nature's authority. By confronting nature and setting it against itself, consciousness comes to regulate and control nature. By Machiavellian intrigue humanity plays one force of nature off against another with the ultimate aim of installing itself as the sovereign power, as master of the universe.[17] "Nature becomes . . . pure Object for man. [sic] It ceases to be recognized as a power for itself; and the theoretical knowledge of its autonomous laws itself appears only as a stratagem for subjecting it to human needs, be it as an object of consumption, or a means of production" (Marx in Schmidt, 1971: 157).

This relationship, which can only be seen as one of exploitation, will, it seems, persist even after the demise of capitalism: for the labor process Marx describes is purported to be universal. Although its content and internal structure—the relations of production—may change, its form remains unaltered: the same elements, that is, "labor power," the "object of labor," and the "instruments of labor" reappear, and laboring remains *the* fundamental human activity, an activity whose *telos* is always the domination of nature. "The labour process . . . is an appropriation of what exists in nature for the requirements of man [*sic*] It is the universal condition for the metabolic interaction [*Stoffwechsel*] between man and nature" (Marx, 1990: 290). Marx famously said little about the form taken by a communist society but it is interesting to read his description of postcapitalist life in "The German Ideology" from the perspective of nature. "[C]ommunist society . . . makes it possible for me to do one thing today and another tomorrow, to hunt in the morning, fish in the after noon, rear cattle in the evening, criticize after dinner, just as I have a mind" (Marx, in Kamenka, 1983: 177). It is not accidental that nature appears even here, in this rare utopian vision, only in a subjugated role, as the object of labor, as something to be ensnared and consumed.

Despite the fact that nature provides the fish that become the angler's quarry, the deer that are the hunter's prey and those rich veins of coal and iron that found industry, it gets precious little credit. Human labor is always the important factor, the real source of all utility and value. The earth may provide "ready-made means of subsistence such as fruits," it may be humanity's "original larder" and "tool house" (Marx, 1990: 285); nature may, "in its original state . . . suppl[y] man [*sic*] with necessaries or means of subsistence ready to hand . . . without any effort on his part" (Marx, 1990: 284), but for all its apparent generosity it becomes valuable only after human ingenuity has improved upon its contributions, reinvesting and reworking its offerings. In a passage that must, from an environmentalist perspective, resonate with dramatic irony, Marx approvingly quotes James Steurt's *Principles of Political Economy*: "The Earth's spontaneous productions being small in quantity, quite independent of man, appear, as it were, to be furnished by Nature, in the same way as a small sum is given to a young man, in order to put him in way of industry, and of making his fortune" (Steurt in Marx, 1990: 285). In other words, Marx compares humanity with an incipient capitalist whom, in order to further his own fortunes, reinvests the capital he acquired through no effort of his own. The fact that he can make such a comparison implies that, although Marx is entirely critical of such economic relations where humans are concerned, he had few qualms about advocating an exploitative relationship between humans and nature. The *works of nature*, those products of labors that provide so many and varied contributions to human well-being, are expropriated and consumed with no thought to the well-being of that which actually produced them.

To summarize, by insisting on anthropic definitions of "labor" and "production," Marx inevitably excludes nature from consideration in and of itself. As Schmidt states, "As far as possible, human labour transforms the in-itself of nature into a for-us" (Schmidt, 1971). Indeed, from a Marxist perspective, the very idea of nature in itself is inherently ambiguous. On the one hand it is the ground of Marxism's materialism—that which precedes and encompasses all human life; on the other hand it is only objectified and valued through the mediation of human labor, that is, only human nature makes it "matter." This ambiguity applies not only to that external nature, whose transformation fuels and facilitates the constant cycle of production and consumption, but for our internal (human) nature too. Humans are at root part of nature and never entirely separable from it (this remains the case even though Marx's insistence on the qualitatively different character of human labor places an irreconcilable tension at the heart of his problematic), but labor is also in part a process of self-realization and transformation, it is a movement that "acts on external nature and changes it" and "simultaneously changes his [the laborer's] own nature." Thus nature is both constantly present behind

Marxism's scenes but hardly, if ever, recognized for what it is in itself. Its presence is both absolutely necessary as a key theoretical concept and, ironically, in another sense it remains entirely absent since we live in a society built out of the constantly reworked products of the past and are ourselves the artificial products of our species' historical labors. To paraphrase Althusser, nature might be regarded as determining "in the last instance" but from Marx's perspective from "the first moment to the last, the lonely hour of the "last instant" never comes."

Nature and the Dialectic, Active and Passive Roles

There is a danger that my references to "nature in itself" might be being misunderstood as an indication that I wish to defend a nondialectical form of materialism. By this I mean that some may regard the critique of the mediating role of labor and the dominance of use-values as a precursor to a straightforward argument for recognizing the intrinsic value of unmediated nature. Chakraverti notes that Marxists have often been concerned to distance themselves from any naive materialism or objectivism about nature, by denying that nature exists "'in itself' without human intervention or mediation" (Chakraverti, 1978: 92). By this I take him to be making an epistemological point, namely that we cannot know what nature is "really" like, since the process of attaining knowledge is itself a form of human activity that transforms the objects of its labor even as it works upon them. In other words we can never have access to an unmediated nature. Alfred Schmidt makes this point as follows: "[t]he *dialectical* element of Marxist materialism does not consist in the denial that matter has its own laws and its own movement (or motion), but in the understanding that matter's laws can only be recognized and appropriately applied by men through the agency of mediating practice" (Schmidt, 1971: 97). I shall have more to say on this issue in the next chapter, but for my present purposes I want to be clear that I am by no means arguing that we have access to an unmediated nature, nature as it "really" is. Nor for that matter can we discover intrinsic values within it that are *independent* of our human relations to it.

My argument is subtler but, I think, just as far-reaching. I fully accept Chakraverti and Schmidt's point about the impossibility of knowing nature as it really is. Nonetheless I want to deny that the process of producing knowledge, or values or any-*thing* for that matter can be understood as just a *social* process, that is, that the dialectic is directed solely by *human* consciousness and driven by *human* labor. That is, I want to challenge the explicit emphasis given to human activities and influences in Marxism's productivism. While Marxism pays lip-service to nature's role in the productive process, the fact that labor, the motive force behind production and the medium through which objectification occurs, is defined as a specifically human activity means that humanity becomes the measure of all things. Such anthropocentrism grants us godlike creative powers to remake the world in our own image. So far as Marxism is concerned, humanity is the Holy Ghost of production, it both inspires and sanctifies all meaningful and worthwhile activity, it is the omniscient and omnipotent specter of the labor process. This is true whether the things produced are goods, knowledge, or values.

Marx's anthropic emphasis bestows a unique position on human labor such that production occurs only at humanity's behest and to fulfill humanity's purposes. Labor is human *activity* and, though not inert, nature is regarded as almost entirely *passive*, its "slumbering" form is "set in motion" by humanity's interventions.[18] Outside of its role as the source of future raw materials and tools the environment appears only rarely in Marx's works and then usually as a constraining influence. In Schmidt's words, human aims are "not just limited by history and society but equally by the structure of matter itself . . . men, [sic] whatever historical condition they live in, see themselves confronted with a world of things which cannot be transcended and which they must appropriate in order to survive" (Schmidt, 1971: 76). Nature is a primeval equivalent of the "slime of history," the all-pervasive medium of our existence from which we try but can never fully manage to escape. Marx does admit that the environment affects social relations in this constraining role. For example, he states that "the quantity of surplus labour will vary according to the natural conditions within which labour is carried on, in particular the fertility of the soil" (Marx, 1990: 648). But, having defined humanity's essence as labor there is a strange inversion in the relationship between humanity and nature, for the more nature does for humanity, the more *active* it is, the more Marx believes we are constrained. "Where nature is too prodigal with her gifts, she "keeps him in hand, like a child in leading-strings'" (Marx, 1990: 649). Humans need to labor in order to "improve" themselves and their environment and once more Marx quotes with approval the bourgeois political economists he otherwise so despises. In the merchant Thomas Mun's words: "The first [natural wealth] as it is most noble and advantageous, so doth it make the people careless, proud and given to all excesses; whereas the second [wealth acquired through labor] enforceth vigilancy, literature, arts, and policy" (Mun in Marx, 1990: 649, fn. 5).[19] Nature it seems can do no right, since even where it is at its most "productive," it is really stunting our development by removing the need to labor, it replaces honest toil with indolence and leisure.[20] When environmental circumstances do make life more difficult "it is the necessity of bringing a natural force under the control of society, of economizing on its energy, of appropriating or subduing it on a large scale by the work of the human hand, that plays the most decisive role" (Marx, 1990: 649). Whatever the natural conditions humanity's aim should always be the same, to dominate nature. There is thus a distinctly Promethean air about Marx's approach. The manifest destiny of humanity is to overcome whatever natural conditions they may find via the "living, shaping fire" (Marx in Schmidt, 1971: 76) of labor and to celebrate their labors as the only ones that make a world worth living in.[21]

The concentration on labor's role in production has two effects where nature's value is concerned. First it means that values are instumentalized—things only become valuable through their being used and transformed, and second that all values are themselves products of human labor. That is, it is only through human intervention that nature becomes valuable and that this value is never anything more than a measure of its potential or actual usefulness to humans. But these claims will not satisfy anyone who wants to speak of nature as being valuable as it is encountered outside of the labor process, and this is precisely what most environmentalists and people who professes to have a concern for nature conservation want to do. And note here that the more narrowly the labor process is drawn, that is, the more it is located within the economy alone, the less one is able to account for nature's apparent value to people even in instrumental terms. For example, some might wish to claim that they regard nature as valuable for the aesthetic enjoyment they get from country walks. But, even such instrumental evaluations as these, are either nonsensical or simply wrong if all values are held to arise economically.[22]

In any event, I want to argue that the value of nature is not reducible to its use for humans, however widely such a notion of utility might be drawn, and that the value of nature is not just a product of human intervention in nature. What is more, I believe it is perfectly possible to hold both these claims without reverting to a naive materialism or an objective or realist theory of value that would regard values as *things* in their own right, or perhaps as being supervenient upon certain kinds of object in the manner that ethical axiologists claim that values supervene on certain properties of natural things like consciousness, sentience, having-a-good-of-one's-own, and so on. (See chapter 1.) One way to do so might be to give due regard to nature's role in the production of things and values. That is, one could widen the definitions of "labor" and "production" so far as to include nature as an *active* partner in the dialectic, a partner that intervenes in its own ways and exerts its own influences at various levels in society—not just through the economy. This might mean that our coming to have certain values might be as much a result of nature's intervention in our lives as our intervention in nature, of its working on us to produce different perceptions and notions of our relationship to our surrounding environment.

Giving nature a more active role in the dialectic would profoundly affect the structure of Marx's thought and, despite the fact that in current circumstances this specificity may be as much a burden as a boon, such suggestions are inevitably extremely controversial. Nonetheless, as the bare minimum required to show that Marxism wants more than to annex environmentalism, it has to engage in serious debate about its anthropocentrism and its instrumentalization of nature, about broadening key concepts, like "labor" and "production," and about the recognition of nature's active participation in the dialectic. These issues become the central grounds for any debate between radical environmentalism and Marxism *if* we are to remain within a productivist paradigm at all. As Martin Ryle states, it "is certainly mistaken to think that 'ecology' can be enlisted, as a 'new social movement,' behind the banner of labourism in the manner assumed by some Left writers . . . the demands of new movements cannot be appended, as ancillary items, to the existing agenda, because they challenge important assumptions [including . . .] the centrality of production" (Ryle, 1988: 17–18).

Productivism and Values: Debates within Marxism

Debates within Marxism about the environment can be seen to fall along a spectrum of opinion. This runs from narrow economistic and Promethean readings of key conceptual terms like "labor," "production," and "dialectic," through those who retain a residual anthropocentrism but are willing to widen the scope of these terms away from an economic focus, and extending to those who are, in some circumstances at least, willing to contemplate even more radical departures from orthodoxy in order to encompass the fundamental issues raised by environmentalism.

Reiner Grundmann has portrayed three types of Marxist response in this debate that we might place along this spectrum. The first is to refuse to make any concessions to environmentalism at all and remain within Marxist orthodoxy. He suggests that Ernest Mandel might belong in this category. The second is to argue, by selectively quoting from his works, that Marx himself was a "Green," "albeit a Green *malgre lui*" (Grundmann, 1991: 103). As the quotations earlier in this chapter show Grundmann seems correct to assume that this is simply a form of "wishful thinking." There is little or no evidence that Marx had any great affinity for nature or, for that matter, nature lovers—quite the opposite.[23] The third strategy he terms "Marxist dissident," because those in this category abandon one or more of the central elements of Marxism, arguing that Green issues cannot be contained within so narrow a framework. Examples of those in this category would be André Gorz or Rudolf Bahro. (See below.) Having established these positions Grundmann goes on to argue that all are wrong in one way or another. Whilst accepting that environmentalism does raise genuine issues that need to be addressed by Marxism, he considers that a "'broader historical materialism' can . . . be revealed by a conceptual reconstruction of the labour process" (Grundmann, 1991: 107) and that this broader conception is capable of delivering an adequate account of ecological problems. Thus, although Grundman does not believe that Marx was any more environmentally aware than his contemporaries he still holds that the framework that Marx created to analyze society is adequate to account for environmental problems and concerns. In this sense he rejects any radical reconstruction of Marx's theoretical problematic.

Grundmann at least has the virtue of refusing to put an environmental gloss on Marx's works but he openly expresses support for the domination of nature that is implicit and explicit in Marx's productivist problematic. Grundmann claims that this domination does not necessarily mean destruction—one can, it seems, be a benevolent dictator who masters the instruments and objects of their labor. "As far as the phrase 'domination of nature' is concerned, there seems nothing wrong with it if it denotes "conscious control'" (Grundmann, 1991: 111). Grundmann seems to suggest that the important issue is not the domination of nature per se but the domination of the dialectic between humanity and nature. On several occasions he suggests an analogy between the mastery of a skill, that is, a practice, like violin playing or teaching, and the mastery of nature. The idea presumably being that the master of her/his craft gets the best performance from their instrument (nature). But this is patently an example of making instrumental rationality the ground for all relationships. It *does* matter whether one lives by manipulatively dominating all around you, by trying to attain "mastery," or by trying to exist *with* them. There is a difference between virtuosity and virtue. Grundmann would obviously not be happy if we applied his understanding of dominance to other humans since this refusal to treat other people as means to one's ends is, after all, what makes the difference between Hobbes and Marx. If we made "efficient mastery" the sole criteria on which we are to judge relationships then, given a labor theory of value, so long as capitalism is efficient in squeezing the most labor out of people then it would also have to be regarded as morally right.

Grundmann's Promethean faith in humanity's destiny and his refusal to take on board anything of environmentalism's critique of instrumental rationality becomes obvious in the manner in which he chooses to define ecological problems. He reduces these to three basic categories: those of pollution, depletion of resources, and population growth. The instrumental bent of these categories is immediately apparent. For example, without any further argument he simply subsumes the destruction of wilderness and species extinction under the heading of "depletion of *resources*." Thus, by the use of a typically narrow anthropic taxonomy, he refuses to even recognize the possibility of there being noninstrumental and nonanthropocentric values. Indeed, having defined matters in this way, Grundmann continues to argue that ecological problems are simply a consequence of a lack of human control over the environment, one caused by insufficient domination rather than resulting from the attempt to dominate. "[W]e term ecological a problem that arises as a consequence of societies dealings with nature. . . . It does not mean that the very fact of dealing with nature (manipulation, domination, harnessing or inducing) is the crucial point, the cause, so to speak, of ecological problems. Ecological problems arise only from *specific* ways of dealing with nature. To repeat my earlier claim: both society's existence in nature and its attempt to dominate nature are compatible; human beings do indeed live in, and dominate nature" (Grundmann, 1991: 113). In the event, Grunmann's "broader historical materialism" and his "conceptual reconstruction of Marx's analysis of the labour process" (Grundmann, 1991: 107) do nothing more than reiterate the instrumental and anthropic concerns of his mentor, though Grundmann bases his faith on the development of new technologies rather than simply the overthrow of capitalism.

Grundmann's optimistic assessment is perhaps indicative of his blinkered attitude to the dialectic. For example, he only mentions in passing the unintended and often environmentally disastrous consequences caused by the application of human technology. He seems to regard these increasingly frequent events as a contingent rather than a necessary feature of human/nature relations. But surely, if we are to take Schmidt and Chakraverti's epistemological point about the dialectic seriously, namely, that we can never know nature in itself, never understand it fully, then we need to beware the danger in molding it to suit our own immediate purposes. The more we force nature to respond to our wishes, the more we dominate it through our control of the dialectic, the less we attend to what nature "says" to us. An obsession with control is actually counterproductive since it ignores the two-way dynamic of the dialectic. Marxists of all people should understand this since this is precisely the principle behind the argument that capitalism's attempt to exert "mastery" over the laborer actually sows the seeds of its own downfall by exacerbating class antagonisms.

A similar point has been taken up in recent debates around what the Marxist James O'Connor (1991) has referred to as the "second contradiction of capitalism." The first contradiction is that induced by capitalist accumulation, which requires a continual decline in labor costs and thus creates a society whose very existence embodies an increasing tension between the rich few and the impoverished many. O'Connor refers to the second contradiction as the "absolute general law of environmental degradation." Put simply, the greater the wealth produced by society the greater those societies' ecological demands and the greater the ensuing environmental degradation. Eventually a point will be reached where ecological systems collapse and thus society too is thrown into turmoil. "Examples of capitalist accumulation impairing or destroying capital's own conditions, hence threatening its own profits . . . are well known. The warming of the atmosphere will eventually destroy people, places, and profits, not to speak of other species life. Acid rain destroys forests, and lakes and buildings and profits alike" (O'Connor, 1992: 189).

The recognition of this second contradiction has been seen by some as a potential way of building bridges between environmentalism and Marxism. (Some Marxists are even willing to argue that it may soon overtake the first contradiction in terms of importance, since capitalism has taken some steps to ameliorate its excesses in the social field [Foster, 1992: 81]). Certainly, at face value, it seems to give nature an active role in the dialectic, in the sense that its activities are at least recognized as producing profound social effects. Yet a number of serious problems remain. Once more, nature's activity is admitted only insofar as it places "constraints" on society. The debate has thus often degenerated into one about the existence and import of natural limits, of what these limits might be and when their transgression is likely to precipitate a crisis. Within this debate radical environmentalists find themselves labeled as neo-Malthusians, proclaiming nature's coming retribution and the immanent end of society in ecological collapse.[24] Against this are ranged the forces of technological Prometheanism. Thus, Grundmann targets the work of Ted Benton, who, he holds, goes too far in his attempt to reconstitute Marxism and wrongly emphasizes that productive processes, like agriculture, have limits that are relatively impervious to human action. Grundmann insists that modern technology has brought about a situation where the possibility exists of overcoming all these natural limitations. "It is ironic that Benton stresses the rigid character of "contextual conditions" and "natural limits" in a world where actual industrial societies explore the possibilities of pushing these barriers further and further back—the substitution of raw materials, development of new synthetic materials, genetic engineering and information technologies being the main examples" (Grundmann, 1991: 108). Benton, quite rightly, rejects this Malthusian label and all such attempts to "equate the ecological perspective with neo-Malthusian conservatism" (Benton, 1989: 52n), arguing that these limits are not universal absolutes that dictate definite social responses but are, though culturally defined, nonetheless real.

However, the problem with focusing on natural limits for radical environmentalism is obvious. O'Connor's second contradiction is unlikely by itself to bring about a crisis in capitalism until the situation for nonhuman nature is depressingly bad, so bad indeed that little of it may remain in a form worth preserving. There is little consolation for environmentalists in Enrique Leff's suggestion that "a generalized ecological crisis, induced by capital accumulation, can bring about a catastrophic crisis of the economic system. . . . Exploited nature [may be] able to accumulate greater anger and ignore the offences of spoilation and extermination with more difficulty than the worst genocide, unleashing the rebellion of natural forces that might become stronger and more uncontrollable than a social revolution" (Leff, 1992: 115). While most radical ecologists want to see the back of capitalism, few would want to celebrate this coming about through the retributive power of global warming or nuclear contamination.

And of course, the very strategies that Grundmann so proudly points to as marks of humanity's technological ability to escape nature's limits—resource switching, genetic engineering, and so on—are postponing the day of reckoning by transforming nature almost beyond recognition and thereby removing much of the reason for valuing it. Although even toxic spoil heaps and sewage-filled rivers may, in the last instance, be regarded as "natural," it is not nature per se that environmentalists value but particular forms of it and they find little value in a world so transformed.

There are other problems with this particular attempt to widen the dialectic. As the introduction argued, environmental degradation has been a feature of all modern (and not a few premodern) societies, not just capitalism (Seymour & Giradet, 1990). As Victor Toledo, commentating on O'Connor's ideas, says, "I am afraid that while trying to relate Marxism and ecology, we *a priori* impute every recognized ecological problem to capitalism, creating commonplaces, not theory, and perpetuating a black tradition of dogmatism" (Toledo, 1992: 86). Of course, it is always possible to make excuses for the fact that those areas of the world that explicitly claimed to embody Marxist ideals have been every bit as destructive ecologically as their capitalist counterparts.[25] But, at the very least, one must admit that the examples set by those states that ranged themselves against capitalism are hardly encouraging for environmentalists.

Toledo raises the further problem of the economic specificity of O'Connor's model, which, he says, "leaves me in a sea of doubts: Is the ecological crisis solely a consequence of an economic contradiction or, on the contrary, does it emerge from a highly complex set of causes—technology, demography, geography, culture, ideology and forms of property? . . . Are we facing a mere crisis of the economic system or a crisis of civilization (which implies a challenge not to an economic rationality, but rather a challenge to a 'mode of life')?" (Toledo, 1992: 85).

It would seem then that there are a number of serious and perhaps insurmountable problems with simply trying to squeeze environmental issues into Marxism's pre-existing framework. These problems remain even if one is willing, like O'Connor, to make substantial additions to that which already exists. The problem lies with Marxism's own incomplete critique of the principles that underlie capitalism and modernity. As Cornelius Castoriadis puts it, "Marx is deeply immersed in the capitalist imaginary. . . . He believes in the centrality of production and economy. He shares in the mythology of 'progress.' He is fully taken by the fantasy of rational mastery of man over nature and himself . . . and the main reason for this is that Marx transforms into a theoretical axiom what is the (unattainable and self-contradictory) practical objective of capitalism: that labour power is (has to become) just a commodity like any other" (Castoriadis, 1992: 204).[26] These are precisely the reasons why those who Grundmann referred to as Marxist dissidents have felt the need to radically reconstruct the productivist problematic in a variety of ways.

Ecology and Productivism from a Dissident Perspective

A number of dissident voices have been raised that, while at least nominally Marxist, depart still further from orthodoxy in an attempt to encompass environmentalism. Perhaps foremost among these are André Gorz and Rudolph Bahro. Bahro, an East German dissident jailed for his criticisms of the pre-unification government, eventually retained only vestiges of his previous Marxism. He adopted a fundamentalist green line that led to his leaving the German Green Party in 1985 because of their becoming embroiled in party politics (Bahro, 1994).[27] His critique of Marxism has tended to focus on issues of practical politics and is partly motivated by his distaste for what he regards as the domination of socialist discourse by the "single issue" of the institutionalized wage struggle. His writings therefore only incidentally touch on the issue of production per se. As Robyn Eckersley suggests, Bahro "is at pains to point out that the challenge of ecological degradation is primarily a cultural and spiritual one and only secondarily an economic one" (Eckersley, 1992: 164).

Gorz, on the other hand, expands upon Marx's theme of the alienation of labor under capitalism, widening the critique to industrial societies in general. He argues that the division of labor in an industrialized economy will always mean that "the "modalities and objectives of work are, to a large extent, determined by necessities over which individuals or groups have relatively little control" (Gorz, 1983: 9). This alienation can never be fully overcome because all complex societies inevitably entail the socialization of work, that is, individuals work to fulfill the expectations and needs of others rather than themselves alone. The *heteronomy* of work continues even under socialism. The control of the means of production by the workers themselves would not alter the fact that there "can never be a complete identity between individuals and their socialized work" (Gorz, 1983: 10). In this sense socialism's "injunction that each individual be completely committed to his or her work and equate it with personal fulfillment . . . mirrors the morality of the bourgeoisie" (Gorz, 1983: 10). It is, Gorz claims, an ethic of "depersonalization." Gorz thus wants to challenge the productivist paradigm by suggesting that "work is not and should not be the center of one's life" (Gorz, 1983: 10). Indeed, he claims, "[f]uture socialism will be post-industrial and anti-productivist, or it will not be" (Gorz, 1985: 3). Gorz suggests that freedom arises only with the *autonomous* control of one's activities. Thus, because heteronomous work cannot be entirely eradicated, "leisure-time" suddenly becomes the focus of debates about the possibility of creating a freer society. This challenge to productivism also inevitably questions the centrality of the working class in Marxist analysis (Gorz, 1983).

This focus on nonproductive activities (in the narrow sense) is doubly important because, Gorz believes, the introduction of new information technologies will drastically reduce the demand for wage labor. "The micro-electronic revolution heralds the abolition of work" (Gorz, 1985: 32). Society thus faces a choice. If more equitable social arrangements that increase the possibilities for autonomous activities are instituted then this freeing up of time might become something to celebrate. If they are not then this technological revolution will simply lead to a greater gap between those with work and those without. "Unlike the mega-technologies of the industrialist era, which were an obstacle to decentralized, community based development, automation is socially ambivalent. The mega-technologies were a one-way street, whereas micro-electronics is a cross-roads" (Gorz, 1985: 29).

Gorz believes that the changes in social relations heralded by new technology also make possible changes in society's relations to nature. Industrial capitalism has an appalling environmental record, but varieties of socialism that continue to place their faith in centralized and large-scale industries, like nuclear power, fare no better. What is required are tools that can be "used and controlled at the level of the neighborhood or community," those that favor "economic autonomy" and that "are not harmful to the environment" (Gorz, 1987: 19). But this kind of technological determinism, which holds that "[t]he institutions and structures of the state are to a large extent determined by the nature and weight of its technologies" (Gorz, 1987: 19), is not actually so very far removed from the productivist orthodoxy he criticizes. His "antiproductivism" actually amounts to little more than changing the model of production from that taking place in large-scale industrial complexes to a decentralized community production of goods under cooperative local control. This decentralized vision may fit with many environmentalist utopias but, for all that, it does not radically alter the Marxian paradigm's emphasis on an economic and anthropic relation to nature. Indeed, although widening the concept of labor somewhat to include those activities that take place outside of the contemporary bounds of the economy, Gorz's concept of the dialectic retains its Promethean character and he makes no concession to those who see nature as anything other than a resource. "Nature is not untouchable. The 'promethean' project of 'mastering' or 'domesticating' nature is not necessarily incompatible with a concern for the environment. . . . The fundamental issue raised by ecology is simply that of knowing: whether the exchanges, which human activity imposes upon or extorts from nature, preserve or carefully manage the stock of non-renewable resources; and whether the destructive affects of production do not exceed the productive ones by depleting renewable resources more quickly than they can regenerate themselves" (Gorz, 1987: 21). It seems then, that nature is still subject to humanity's extortion racket, its protection dependent upon its continuing to pay under duress what humanity considers its dues and so ensure its dreams of a life of leisure.

However unorthodox Gorz may be from a Marxist perspective, from the point of view of radical ecology he retains a form of productivism that is in no way less anthropocentric. While being less sanguine about the need for humans to work, he is no less convinced of the need to *manage* nature. These close ties between productivism and instrumentalism might lead us to question the very notion that productivism can in any way aid radical ecology. It is to just such a critique I now turn.

Baudrillard and the Mirror of Production

To summarize so far: Those reductivist positions that hold rigidly to the base/superstructure model have overemphasized the determining role of the economy (narrowly conceived). Recognizing this and rather than trying to economize all aspects of society, Althusser introduces what amounts to separate paradigms of production for the political, ideological, and theoretical spheres. All that they have in common is that they are transformative and socially mediated processes that entail the expenditure of human labor. But, while introducing different forms of production may make theory less economistic, the question still remains as to why one should refer to certain practices as *productive* at all? Can "production" be an appropriate or adequate metaphor to signify the multiplicity of practices found in all societies? Where precisely does the analogy between economic and, say, theoretical practices lie? Isn't it simply bizarre to think that all human activity could be accommodated within a framework of "production"?

This question has concerned Marxists like Habermas, who asks how "the paradigmatic activity-type of labor or the making of products [. . . can be] related to all the other cultural forms of expression of subjects capable of speech and action" (Habermas, 1987: 79).[28] In particular, Habermas wonders how the communicative activities that lie at the heart of his social and ethical philosophy could possibly be conceived of in terms of labor. Habermas' point can be extended, for it isn't only communicative activity that brings the concept of production into question. Such things as taking country walks, mountaineering, playing chess, or saving someone's life don't appear to fit into categories of *productive* practice without an immense and distorting effort, a theoretical contrivance that stretches credulity. Of course, one could, if one so wished, argue that these practices were productive of enjoyment or fitness, but to do so would be to utilize a very etiolated conception of production and of labor.

The problem is that the further one moves away from the *economic* archetype of production, the less the analogy seems to hold and the more it loses its Marxist specificity. As Seidman states, "[t]he category of 'productive' activity either expands to include virtually all human practices, in which case it is useless as a conceptual strategy, or it narrows arbitrarily to economic labouring activity" (Seidman, 1992: 57). Thus Marxists have a problem with reading "production" in terms of objectification since this might actually undermine that emphasis on the primacy of the economic sphere that is so fundamental to the character of Marxism in general. On the other hand, although Althusser's metaphysics manages to uphold, to a degree, the possibility of a holistic but antireductionist strategy, his continued reliance upon the economic metaphors of "production" and "labor" as the *only* forms of social mediation seems to constrain as much as enable our thinking of certain practices and of modernity in general. The question is whether we should abandon the productivist ship altogether or whether it is salvageable. Jean Baudrillard has been one of the foremost critics of productivism's seaworthiness and hence his analysis provides an obvious starting point from which to discuss these issues.

Baudrillard considerably widens the debate about the economic reductivism inherent in the productivist paradigm. In *The Mirror of Production* he argues that productivism of *any* form is indelibly stamped with an economic rationality. This work is described by Baudrillard's English translator, Mark Poster, as a marshalling of his earlier work "for a systematic critique of Marxism" (Poster in Baudrillard, 1975: 1). But the effects of this critique are felt on a wide variety of theoretical problematics. Not only is the work a critique of Marx's own writings but it extends to all those philosophies that have emerged from Marxism and carry over the metaphor of production into their new problematics. This might include, for example, as Poster indicates, such "postmodern" thinkers as Gilles Deleuze and Felix Guattari (1990; 1998), who refer to the *production* of desire. Baudrillard considers that retaining the production metaphor is indicative of a conceptual conservatism that is both theoretically and practically stifling. "Production" becomes a singular and universal "sign" that rules our thought and actions. By contrast, Baudrillard seeks the reinstatement of a symbolic multiplicity that will, he claims, subvert modernist values (Best & Kellner, 1991: 115–16).

Baudrillard argues that the origins of the problems that beset productivism lie in Marx's critique of his contemporary political economists. Despite all appearances, this critique was simply not radical enough. Marx only succeeded in replacing current myths with "a similar fiction, a similar naturalization—another wholly arbitrary convention, a simulation model bound to *code* all human material and every contingency of desire and exchange in terms of value, finality and production" (Baudrillard, 1975: 18–19). Marx failed to adequately deconstruct and hence escape the constraints imposed by the categories of production and labor. As Poster puts it "Marx's theory of historical materialism . . . is too conservative, too rooted in the assumptions of political economy" (Poster in Baudrillard, 1975: 1). The sign of "production" becomes reified as an objective and essential process necessary to all human social being. Under this tyranny of the sign humans come to be defined as productive animals, a definition that severely constrains our theoretical understanding of human activity and those activities themselves. In effect Marx makes "human labor" the new universal human essence.

Baudrillard questions the very possibility of extending a paradigm of production to cover the totality of human practices in all their different forms: "A specter haunts the revolutionary imagination: the phantom of production. Everywhere it sustains an unbridled romanticism of productivity. The critical theory of the *mode* of production [i.e., Marxism] does not touch the *principle* of production. All the concepts it articulates describe only the dialectical and historical genealogy of the *contents* of production, leaving production as a *form* intact" (Baudrillard, 1975: 17). In other words, Baudrillard's point about productivism being irretrievably contaminated by its modernist heritage is analogous to my own critique of contemporary ethical and social theories in chapters 1 and 2. Although the content of the theory (of production) changes from Marx to Althusser to Deleuze and Guattari the form remains relatively intact and its explanatory power resides in its very conformity to its modernist assumptions. For this reason they are all, says Baudrillard "locked into a speculative dead end" (Baudrillard, 1992: 103).

Marx's reification of labor power as "the fundamental human potential" (Baudrillard, 1992: 104) is indicative of his retention of the structure of the very political economists he wishes to criticize, that is, of a mode of thinking that epitomizes and drives the process of rationalization. (See chapter 1.) "More deeply than in the fiction of individuals freely selling their labour power in the market [the fiction that forms the heart of Marx's critique of political economy], the system is rooted in the identification of individuals with their labor power and with their acts of 'transforming nature according to human ends'" (Baudrillard, 1992: 104). Ironically, Marx and Marxism are, says Baudrillard, complicit in "censoring" more radical critiques of capitalism, since they retain the very same instrumental conceptions of human nature and rationality as those that underpin modernism. Marxism is guilty of an "aberrant sanctification of work," it reiterates and elevates the Protestant work ethic to new heights; work (the means) becomes an end in itself as Marxism "generalizes the economic mode of rationality over the entire expanse of human history" (Baudrillard, 1992: 105).[29]

Marx's "work ethic" is no less reductive and anthropocentric than the evaluative axiologies of other economists, it reduces all values to "use-values," and to the concrete modes of labor that create such values. An object's value thus becomes entirely dependent upon its ability to fulfill *human* needs and upon the amount of *human labor* expended upon producing it. Of course, the emphasis on use-values is motivated by the need to provide a firm foundation for a critique of capitalism. In this way Marx reasserts the origins of all values in human needs and labor rather than in systems of exchange. Use-values are the *reality* behind the phenomenal form of value's *appearance* as exchange-value (Marx, 1990: 128). However, as Baudrillard points out, to emphasize use-values is not to escape from capitalism's economistic paradigm because "[f]ar from designating a realm beyond political economy, use value is only the horizon of exchange value" (Baudrillard, 1992: 99). Contra Marx, Baudrillard insists on the dependence of use-values upon systems of exchange-value. Our coming to see use-values as somehow expressing "natural" needs is actually economic rationality's greatest accomplishment, for such needs are actually manufactured within and by the capitalist system. "The definition of products as useful and as responding to needs is the most accomplished, most internalized expression of abstract economic exchange: it is its subjective closure" (Baudrillard, 1992: 100). In other words "exchange-value" and "use-value" are two poles of a dialectic that work together to ensure that everything comes to be economized, that all becomes subservient to an economic rationality.[30] This economization of the world is, as previous chapters have illustrated, a primary cause and effect of the rationalization process.

The effects of this rationalization are nowhere felt more strongly than in terms of "nature," which finds itself reduced to no more than a means to fulfilling these created needs. To the extent that it serves no "useful" human purpose, nature is literally valueless. And here the connection between use-value and exchange-value is again made plain. For, once naturebecomes available for *use*, as the source of those raw materials that labor will work upon, it acquires value; but this value is not its own and exists only for so long as it remains in use within the economic sphere. This value can rise, fall, or disappear altogether as circumstances and needs change—in other words it is an *abstract* value, a value that can be taken from it and given to something else, that can be exchanged and serve to signify nature's own exchange from hand to hand. Both use-value and exchange-value are retained only so long as the "product" is caught up in the continual processes of production and consumption. In Marx's words, anything "which is not active in the labour process is useless, it falls prey to the destructive power of natural processes. Iron rusts; wood rots. Yarn with which we neither weave nor knit is cotton wasted. Living labour must seize on these things, awaken them from the dead, change them from merely possible into real and effective use-values" (Marx, 1990: 289). In this economistic problematic, value is present only due to a thing's presence in a series of constant transformations that leave nothing as it is but work upon things to alter them often beyond recognition. This is, of course, patently at odds with attempts to value and conserve nature as it is, that is, to allow at least some parts of it to remain untransformed by human activity.

Both use-value and exchange-value are assigned by an economic system that depends for its very existence on the constant incorporation and transformation of nature. Neither kind of value can serve as a basis for the recognition of those values that are not regarded as abstractable or exchangeable (an issue that will crop up again over arguments about nature's intrinsic value in the next chapter). To this extent Marxism's productivist system of values both expresses and furthers the processes of modernization and rationalization; it is intimately involved in that same process of valuation that leads to the disenchantment and destruction of the natural world in an ecstatic frenzy of production and consumption.[31]

Baudrillard's critique is compelling in many ways. In many respects he extends Marx's own critique of commodification and turns it against itself. For Baudrillard, as for Marx, "political economy is this immense transmutation of all values . . . into economic exchange values" (Baudrillard, 1981: 113). Subjects and objects alike find themselves caught up in a process of rationalization that reduces all qualitative differences to the singular and quantitative logic of capital. But for Baudrillard people's needs too are manufactured and fetishized. Indeed, such is the dominance of the abstract sign over people and things that, in Steven Best's words, use-values and indeed all attempts at representation, "no longer refer beyond themselves to an existing [independently] knowable world. . . . The real, for all intents and purposes is vanquished when an independent object world is assimilated to and defined by artificial codes and simulation models" (Best, 1995: 54). The absolute domination of our contemporary (post)modern world by systems of symbolic exchange undermines the very idea of "reality," since our lived reality is always already encoded in symbols whose value and meaning resides in their exchangeability and not by their maintaining a fixed reference to external reality. "The symbolic is neither a concept, an agency, a category, nor a 'structure,' but an act of exchange and *a social relation which puts an end to the real*" (Baudrillard, 1993: 133). This situation is what Baudrillard later refers to as "hyperreality."

But while this position condemns Marx's utilitarian ethos as, in many ways, complicit in modernity's abstract rationalization of the world, it also has profound implications for environmentalism. In particular it raises important questions as to the very possibility of speaking of nature in contemporary society. If our conceptual grasp of "reality" has become so tenuous as to float free in a surreal cyclone of constant symbolic exchange, how can we even begin to speak of nature *in itself*? For Baudrillard, "nature" too exists only as a "simulacrum," entirely divorced from any reference to an unmediated extrasymbolic world. Environmentalism is simply part of that "thriving nostalgia for the natural referent of the sign, [which survives] despite several revolutions which have begun to shatter this configuration (such as the revolution of production when signs ceased to refer to a nature and referred instead to the law of exchange)" (Baudrillard, 1993: 50). Ecological crises simply hold out a forlorn promise of the "return of a lost referentiality to the economic code, and will give the principle of production a gravity which evaded it" (Baudrillard, 1993: 32). In other words, because Baudrillard believes environmentalism to proffer a doctrine of "natural limits," an ecological crisis seems to reopen the possibility of grounding the symbolic order in "reality." For this reason, Baudrillard claims, people are happy to blame the contradictions of capitalism for the environmental crisis because the "comforting" thought that we can identify a subject (capitalism) to blame for our predicament is better than accepting the "terror of the simulacrum," of a situation where reality is actually unfathomable. It is impossible to tell "how much truth and how much simulacrum there may be in this crisis" (Baudrillard, 1993: 32). Baudrillard is sensitive to the anthropocentrism implied in the productivist model. He asks why Marx never doubts that humans "begin to distinguish themselves from animals as soon as they *produce* their means of subsistence"?[32] The really interesting question is not *whether* or *how* humans can be distinguished form the rest of nature but *why* "must man's [sic] vocation be always to distinguish himself from animals?"[33] Yet Baudrillard's model implies an anthropic Prometheanism of a different (symbolic) order. Somehow the use and exchange of symbols is supposed to create and control the world, rather than the use and exchange of labor. Baudrillard differs from Marxism here in only two ways. First, in claiming that the dialectic becomes mediated *symbolically*, in a process of simulation, rather than through the labor process. Second, in claiming that the "relative" autonomy of the economic sphere from nature is progressively transformed into the complete autonomy of the symbolic sphere.[34] This symbolic dialectic may no longer be under conscious human control, but nature fares no better; the cycle of production, reproduction, and consumption has become so intense, that all has become artificial; nothing at all remains outside a symbolic order that creates and masters its own universe.

But this one-sided approach to the dialectic simply reintroduces the problem of economism in a new way. We now face a variety of symbolic reductivism where the meaning and values of things are wholly determined by their use in a constantly shifting symbolic order. "Society" and "nature" alike are nothing more than signifiers without a signified, mere tokens of exchange within a semiotic structure. And here again Baudrillard mirrors productivism himself in suffering from an anthropic forgetfulness about nature's own active and creative engagement in the dialectic. The fact that we can't *know* nature *in itself* does not mean that nature plays no role in the production of knowledge. It is true that the state of nature in itself, and the reality of the crisis, may be unfathomable in any absolute sense; such indeterminancy is guaranteed because the only access we can have to nature is mediated dialectically through social systems. But to move from a position of nature's *indeterminancy* to one of nature's *unreality* is to entirely neglect the many and varied ways that nature intrudes into the processes of production and simulation.

I shall return to the question of nature's "agency" in the following chapter. However, it is clear that, despite his accurate critique of the role of productivism in exacerbating the processes of rationalization and economization in modernity, Baudrillard's own problematic is, if anything, even less aware of "nature" in its own right. The exclusion and diminution of nature does not then stem from productivism per se, though its economistic forms are anathema to any radical environmentalism. One cannot, as Baudrillard does, simply excise the term "production" from the extremely variable problematics in which it has different roles and different meanings and claim that they are all inevitably tainted by past economic associations. This is not to deny that a word's past associations have had a profound influence on the way its meaning has been interpreted, but one must take each case as it occurs. A radical environmentalism would have different interests at heart and would use terms differently, and a change of *use* inevitably implies a change of *meaning* (Wittgenstein, 1988). A different, (theoretical) context necessarily implies a displacement of concept (a connection made in Derrida's term "*différance*").[35] Many theoreticians, as we shall see, use "production" in a manner that does not look for its archetype in Marx's later economism at all. A productivism that could give scope for nature's participation in the dialectic may yet prove more fruitful.

This novel form of productivism might help conceptualize our current predicament and provide a framework for rethinking our natural relations. But this productivism would have to avoid imposing a totalizing and reductive schema (such as economism) onto the dialectic between humanity and nature; instead it must emphasize the variety and multiplicity of such relations rather than stamping them with a unitary code or privileging one axis over all others. It would have to accept that "objectification" occurs in all manner of ways, not just through the labor process, and that nature too plays an active participatory role in the production of knowledge and values. In this sense a reformulated productivism must be aware of the influences of different social and natural contexts on the production process and its own influence on those contexts. In other words, there are different "modalities of production" in which nature enters the dialectic in different ways and with differing effects.

4

To Speak of Trees: Social Constructivism, Environmental Values, and the Futures of Radical Ecology

The last chapter surveyed a variety of productivist accounts of social and natural relations beginning with Marx's own. From these, it becomes clear that productivism, especially in its more economistic forms, has frequently served to promulgate an arrogant and anthropocentric attitude toward nonhuman nature. Nature is valuable only insofar as it is transformed by human labor and fulfils human needs. Baudrillard's critique, though drawn too widely, rightly stresses how both use and exchange values are indicative of that process of rationalization that increasingly dominates all aspects of the modern world. The value and meaning of all things becomes located in increasingly complex, abstract, and self-referential processes. We are swept along, with or without our consent, into an increasingly artificial and disenchanted world where all is subject to managerial control and reduced to a heartless calculus of "efficiency." Unfortunately, for all his insights into this process, Baudrillard proffers nothing but more of the same.

Initially, at least, productivism seems to offer little to radical environmentalism. Its theoretical usefulness is compromised by its literal and metaphorical focus on human productive activities as creating all meaning and value. It seemingly makes our knowledge of nature, and our attitudes toward it, entirely dependent upon the mode of its appropriation within differing social formations. This relativization of nature contrasts strongly with the position of those environmentalists who want to speak of nature's "intrinsic" value. Indeed, the power and the promise of radical ecology has been seen, by its supporters and detractors alike, to lie in its claims to speak on behalf of a natural world threatened by human excesses. Yet to speak of trees *as trees* or nature as something worthy of respect in itself has appeared increasingly difficult in the light of productivist accounts.

This difficulty has been highlighted in recent debates between "deep ecologists" and "social constructivists." Deep ecology has been loath to take seriously constructivism's insights into the influences of differing social circumstances on our conceptions of nature. Instead, it has retreated into forms of philosophical objectivism and biological reductionism in order to defend nature's intrinsic value. These debates raise many important issues about our understanding and evaluation of terms like "wilderness," "ecosystem," "biological diversity," and so on, and about the potential conflicts that can arise in any discussion of nature's value. Yet such antagonisms are ultimately self-defeating; deep ecology actually has much to gain from constructivism and can add a new dimension to constructivism's own critique of current ideologies.

To Speak of Trees: A Brechtian Prologue

> Ah, what an age it is
> When to speak of trees is almost a crime.
> For it is a kind of silence about injustice. (Brecht, 1959: 173–77)

It is perhaps ironic that, in an age of environmental crisis Brecht's lines have acquired a poetic ambiguity they previously lacked. In 1938 as he wrote *An Die Nachgeborenen* (to those who are born later), time spent contemplating trees might easily have been regarded as a luxury one could ill afford, a criminal distraction from the all too pressing needs of his age. Yet today there is almost universal agreement that trees need to be at the very center of our political debates, that the fate of the forests and other environmental issues are inextricably tied up with the future of human society and of social justice. The question is not whether we *should* speak of trees but of *how* we should do so.

In the context of environmental ethics this newfound ambiguity in Brecht's lines marks an apparently insurmountable divide between an instrumental anthropocentrism and the biocentric approaches favored by deep ecologists.[1] The former position regards nature primarily as a resource to be distributed according to principles of human justice or need—the latter sees nature as having moral standing in itself, as being of "intrinsic value" or "inherent worth."[2] While all would agree that nature is a prerequisite for social life, to speak of nature as being valuable in itself is still symptomatic, for many, of a moral failure to prioritize the immediacy of human needs over our maltreatment of the environment. From an anthropocentric perspective, values are human creations and as such must be subject to changing human requirements. Deep ecologists and radical environmentalists who insist on speaking of trees per se risk being regarded by those on the political left and right as almost criminally misanthropic.

This debate is played out in a wide variety of discourses that express radically different cultural orientations to nature. For example, to the Canadian lumber industry, trees are just a potential resource, so many board feet or cubic meters of timber, waiting to be used.[3] So far as the Forest Alliance of British Columbia (F.A.B.C.), an organization representing commercial logging interests, is concerned, "British Columbia's forests are its principal natural resource" (F.A.B.C., 1997). It claims that logging companies are merely filling the increasing world demand for "forest products", a demand that is forecast to rise by 86 million cubic meters annually, an amount roughly equivalent to British Columbia's total annual output. Canada's forests must be "managed" for (economic) sustainability under the auspices of schemes like Vancouver Island Resource Targets (VIRT). This plan, whose main purpose is "to enhance timber supply and forest employment," designates "most of the remaining ancient forests as High Intensity Areas (HIA's) for logging" (Poanes, 1997).

To those individuals and organizations ranged against the loggers, these attacks on the "magic groves" and "temple giant old growth forests," cannot be justified. They are a form of "cultural rape" practiced by a dysfunctional and ecocidal society (Morgan, 1994). Critics point out that "of over 170 watersheds on Vancouver Island only nine remain intact," and, despite a much vaunted Forests Practices Code, 92 percent of logging in the province is by clear-cutting (Advocates for the Environment, 1997: 4–5). The choice is between "clearcuts, tree farms and big stumps, or grizzlies, wild salmon and big trees." Canada's "great bear wilderness" with its "pristine valleys" is "too precious to destroy" (Wilderness Committee, 1997).

This is not to say that protesters are concerned only with the damage done to nature; they aren't. They continuously strive to make connections between human well-being and environmental health and point to the illogicality of such overexploitation even within the logging industry's own terms of reference. Loggers will inevitably make themselves unemployed by removing the source of their livelihood.[4] The First Nations people, the original human inhabitants of the forests, also emphasize not only their spiritual and historical connections with the land but their financial interests in maintaining its current state.[5] But the bottom line is that, so far as many protesters and First Nations people are concerned, at least some areas of forest should not to be treated as resources at all, but are due respect in their own right.

These conflicting discourses can be found everywhere. Mahdav Gadgil and Ramachandra Guha list the various "ideologies of environmentalism," including, "crusading Gandhians," "ecological Marxists," "appropriate technologists," "scientific conservationists," and "wilderness enthusiasts," that have sought to influence debates over environmental degradation in India. Each group takes a different position over the social and moral status of "nature" (Gadgil & Guha, 1995). Gadgil and Guha themselves emphasize nature's status as a resource rather than its import in its own right. Forests they suggest perform five major functions; maintenance of soil and water regimes, conservation of biological and genetic diversity, production of biomass for subsistence, and production of timber and nontimber biomass for commercial purposes. With the partial exception of the preservation of diversity, which is itself justified largely by the possibility of discovering as yet unknown resources to fulfil future human needs, these are all anthropocentric categories.

This anthropocentric and instrumental approach is certainly understandable given the relative resource deprivation of most of India's population. However, it may itself be an indication of the pervasive influence of a modernist Western ideology that emphasizes a particular "rational" and technological solution to so-called underdevelopment. It is interesting that most of the indigenous "ecosystem people," who, Gadgil and Guha note, "know of a vast number of uses of the many plants of tropical rain forests" and whose "practices have . . . been moulded to working closely with nature," tend themselves to regard nature as much more than simply a resource (Gadgil & Guha, 1995: 141). In other words, a noninstrumental attitude toward nature is not simply another imported neocolonial discourse, as the authors sometimes suggest, but a constitutive part of the centuries of dialectical coexistence between native Indians and their natural localities. For example, while their educated upper-caste neighbors regard much of their belief system as pure superstition, the lower-caste untouchables of Kerala believe certain trees are both "sacred" and "fearful." The wild naturally occurring sacred groves (*kaavus*) that embody the principles of "life-force" (*shakti*) and "fault" (*doosham*) are regarded as temples by the lower castes because of the powers associated with them (Uchiyamada, 1998: 179). While this has not given Kerala's forests complete protection, it has meant that, despite agricultural pressures, the land remains "densely greened with trees and groves" (Uchiyamada, 1998: 196).[6]

Vandana Shiva notes how a spiritual conception of nature has been profoundly influential in the success of the Chipko movement in Garhwal Himalaya. The Chipko movement is a women-centered campaign informed by Ghandian tactics that seeks to save trees from felling by hugging them, thus placing the protesters own body between the tree and the saw. Shiva describes "two paradigms of forestry—one life-enhancing, the other life-destroying. The life-enhancing paradigm emerges from the forest and the feminine principle; the life-destroying one from the factory and the market" (Shiva, 1994: 76). In this way, rather than simply describing cultural practices in (modern) functional and instrumental terms, which emphasize the "usefulness" of otherwise "irrational" belief systems, Shiva's references to the "feminine principle" echo the beliefs and practices of the protesters themselves.

Understandings of what trees are and what they represent obviously change from culture to culture, from time to time, and even from person to person. But even in the most instrumentally orientated cultures some people harbor different attitudes to trees and nature. The naturalist W. H. Hudson told a story about the owner of a large estate whose "greatest pleasure was to sit out of doors of an evening in sight of the grand old trees of his park, and before going in he would walk round to visit them, one by one and, resting his hand on the bark he would whisper a good-night. He was convinced . . . that they had intelligent souls and knew and encouraged his devotion" (Hudson in Thomas, 1984: 192). Yet, only two centuries before Hudson's landowner, a dictionary could speak of the "savage" (from the word *silva*—a wood) woodlands as "'dreadful,' 'gloomy,' 'wild,' 'desert,' 'uncouth,' 'melancholy,'" and so on (Thomas, 1984: 194), they represented an alien space that humans entered at their peril. This perilous "wildwood" still survives in the fictional forests of fairy tales, of Tolkien's Mirkwood, the *Blair Witch Project*, or, in quite a different way, in the woods in Shakespeare's *Midsummer Night's Dream*.

There are innumerable different ways of relating to trees. Simon Schama, emphasizing the strength of the links that bind society and nature together, recounts a changing history of representations of the forest through the woodland sacrifices of the ancient Germanic tribes, to the Greenwood myths of Robin Hood and the religious symbolism of the "verdant cross." In the introduction to her collection *The Social Life of Trees* Laura Rival points out that today "[w]ith modernity taken to be the primary cause of deforestation, trees further symbolise anti-modern (i.e. anti-urban, anti-consumerist and anti-industrial) values" (Rival, 1998: 16). The treetop protests of antiroads protesters in Britain, the antilogging protests on the West Coast of North America, and the Chipko movement in India can all be seen in this light. And it isn't just committed environmentalists who feel this affinity for trees as links to the past. This is illustrated by the community defense of the 240-year-old Sweet Chestnut Tree on George Green, Wanstead, East London. Defying attempts to push through the new M11 link road, two hundred local residents and protesters battled in the early hours of the morning to defend the tree against an army of some four hundred police. The tree was eventually felled on December 7, 1993 (Evans, 1998: 28–31).

In other words, there are plenty of examples of people regarding trees with a respect that has nothing to do with their instrumental value. They regard them as much more than a potential door or toilet-seat. However, these examples might cut both ways, for if they show anything at all, they show just how dependent our conception of nature is on our cultural circumstances. And to social constructivists this "cultural relativism" suggests not that trees can be spoken of in themselves but that they always acquire their meaning and value *through* culture. In other words, nature is socially constructed and there are no such things as intrinsic values.

The present chapter tries to negotiate the apparent contradiction in holding that trees might be regarded as valuable in their own right and that values are culturally relative. However, my intent is not to reiterate the debate over intrinsic value in its standard philosophical form, but to reconsider it in the light of sociotheoretical arguments about the social construction of nature. The rationale behind this shift in emphasis is twofold. First, to try to avoid the tendency for philosophical debates to fall back into protracted, but not very productive, arguments about moral realism, often seen as the only defense against an encroaching anthropocentric relativism. Second, and more importantly, to engage with recent polemics between certain deep ecologists who are overly pessimistic about the impact of social constructivism on biocentric evaluations of nature and certain constructivists who are overly dismissive of deep ecology's biocentric claims.

Social Constructivism's Challenge to Nature

"Social constructivism" encompasses a variety of theses, some stronger than others. Drawn widely it might include all those theoretical frameworks that emphasize the determining influence of social relations on the production of knowledge, values, beliefs or behavior, that is, on our "lived" realities. In this broad sense it might include even an overtly materialist discourse like Marxism that regards ideologies as products of particular economic relations. Drawn more narrowly, constructivism is more usually associated with phenomenological, ethnomethodological, or postmodern perspectives focusing on the role of cultural, institutional, symbolic, or linguistic systems in delimiting and molding our perceptions of the world.[7] The focus here tends to be on how people assign meaning to their world (Hannigan, 1995: 32; Best, 1989).

Rather than taking claims about the social and natural world at face value, social constructivism emphasizes the influence of different histories, traditions, social practices, power relations, and so forth on the models we produce and utilize. In this way constructivism problematizes any claims made by discourses to provide neutral or objective accounts, insisting that, to some extent, *all* accounts necessarily reflect the particularity of their origins in given social circumstances. This social relativism even extends to our representations of nature, usually regarded in modern occidental society as an objective realm set apart from the vagaries of culture (Wilson, 1992). A burgeoning literature in history, anthropology, and sociology sets out to show how nature has been envisaged in widely divergent manners in different periods and cultures.[8]

Almost inevitably the conclusion of such studies is "that there is no singular "nature" as such, only a diversity of "contested" natures; and that each such nature is constructed through a variety of socio-cultural processes" (Macnaghton & Urry, 1998: 1). This idea of a contested nature looms large in many recent texts in cultural theory and sociology, for example, William Cronon's edited collection *Uncommon Ground: Toward Reinventing Nature*, which emphasizes the cultural construction of nature.[9] In Cronon's words, "[n]ature will *always* be contested terrain" (Cronon, 1994: 52).

The implicit (and explicit) relativism entailed by social constructivism, which for some authors subverts even the natural sciences' claims to encapsulate nature as it "really" is, has massive implications for those who want to speak *for* nature (Latour & Woolgar, 1986). In his account of constructivism in the history of science, Jan Golinski (1998: 6) states that constructivism "draws attention to the central notion that scientific knowledge is a human creation, made with available material resources, rather than simply the revelation of a natural order that is pre-given and independent of human action." This relativization of scientific knowledge proves too much for many scientists. In a recent response to what he refers to as "postmodern deconstructivism" conservation biologist Michael Soulé, attacks social constructivism as an academic "fad" that nonetheless provides a "rhetoric that justifies further degradation of wildlands for the sake of economic development" (Soulé & Lease, 1995: xv). He believes that this "form of intellectual and social relativism can be just as destructive to nature as bulldozers and chainsaws" (Soulé & Lease, 1995: xv).[10]

Constructivism obviously poses particular problems for deep ecology. If social constructivism rejects all claims to grasp nature in pure or unadulterated form, if we cannot claim to have a discourse that truly represents nature in itself, this inevitably problematizes any claims that nature might have intrinsic value. In deep ecologist George Sessions' words, the acceptance of constructivist doctrines would "make hash out of the claims of deep ecologists and conservation biologists that the independent reality and integrity of the Earth's wild ecosystems, biodiversity and evolutionary processes have *intrinsic* value and must be protected for their own sakes" (Sessions, 1995: 15). Social constructivism's "epistemological doctrines" undermine biocentrism and "conveniently serve to bolster a particular ideology and the anthropocentric social/political agendas of Marxists and postmodernists" (Sessions, 1995: 15).

Sessions' assessment seems to concord both with those who reject social constructivism, like Soulé, and those "constructivists" (in the broad sense) critical of deep ecology's claims to speak for nature. From Sessions' and Soulé's perspective, deep ecology and environmentalism in general is faced with an apparent amalgam of Marxists and deconstructivists who in emphasizing the "social history of nature" thereby translate talk of nature's intrinsic value into questions about the value of nature within and for specific societies. By focusing on the contested status of nature the ethical debate is shifted from one of nature's intrinsic value to its instrumental value as a good to be distributed according to principles of *social justice*. Thus for example, Marxists like David Pepper are "uncomfortable" with suggestions that "we should protect and respect nature for its 'intrinsic worth,' *whatever that might be*, rather than its worth for all people" (Pepper, 1993: xi, my emphasis). He wishes to move the debate from deep ecology to social justice. John O'Neill rejects deep ecology in favor of an Aristotelian approach centered on "human well-being" where nature is reconceptualized in the light of "a set of objective goods a person might *possess*" (O'Neill, 1993: 3, my emphasis). Peter Dickens believes that "contemporary environmentalism is characterized by a profound failure to understand their relations with nature" (Dickens, 1996: 149). His preferred solution is a strategy that "combines environmental politics with questions of power and social justice" and claims that these are "often forgotten by elements of the environmental movement" (Dickens, 1996: 16). These Marxist criticisms are reinforced by deconstructivist approaches from cultural theorists. For example, Andrew Ross castigates the "nature nerds" who presume to find authentic knowledge or values by "hanging out among trees and rivers." "There are no 'laws' in nature, only in society, because laws are made by us and can therefore only be changed by us." "Environmentalists," he claims, "are often oblivious to such social milieux" (Ross, 1994: 15). Keith Tester similarly berates the "naturalist assumptions" of those involved in debates about animal rights who are "fetishistically upholding obligations which are *made* and not *found*" (Tester, 1991: 195).

However, from a deep ecological perspective this sociocultural colonization of the natural world typifies that anthropocentric hubris that allows blinkered humanists to regard their theoretical problematics as complete without regard for nature's own being (verb). Nature's human progeny haughtily dismiss its nascent potential declaring it to be their own offspring—an invention constantly remade in our own image. In this sense, social constructivism brings with it a fear about the cultural appropriation and thence dissolution of nature in current theoretical debates.

It seems that the debate over social constructivism reaches something of an impasse. While one side regards all talk of nature as merely an expression of social mores and structures, others see nature as an ahistorical and universal ground for society. While the latter regard at least some discourses (usually those of the natural sciences) as transparent windows onto the natural world—words that tell it as it is, constructivists see all discourses as opaque products of particular social circumstances—each language-game hiding as much as it reveals, selectively sculpting natural forms to suit social purposes.

Varieties of Constructivism

Perhaps this impasse is most easily understood in Klaus Eder's terms. Eder suggests that we regard social constructivism as a reaction against, and critique of, those naturalistic explanations that sought to explain societal evolution and reproduction as a continuation of natural processes. Here one might think of Herbert Spencer's "evolutionary" account of social development or more recently of the attempt by sociobiologists to explain culture in terms of genetics, an account recently extended to include our evaluations of nature (Kellert & Wilson, 1993). In this sense "[t]he relationship between nature and society can be conceived of theoretically in two mutually exclusive ways: as a natural constitution of society or as a social construction of nature" (Eder, 1996: 7). We can describe either a "natural history of society" or a "social history of nature" (Eder, 1996: 19). Taken to its logical conclusion, social constructivism comes to view even nature itself as a product of social relations rather than an unchanging and objective external reality.

However, one of the reasons for the impasse is that both sides of the debate have tended to simplify and misconstrue their opponents' positions, thereby replacing a complex spectrum of opinion with a binary opposition that best suits their polemical purposes. For example, in making an enemy out of social constructivism per se, Sessions conflates very different philosophies, claiming that both "Marxists and postmoderists alike" (Sessions, 1995: 15) propound a "sociological and relativistic analysis of truth" (Sessions, 1995: 15). But this blanket rejection of social constructivism fails to recognize the real disagreements that exist among those referred to as constructivists. Marxists have heavily criticized postmodern epistemologies precisely because they have regarded them as (too) relativistic (Norris, 1990; Callinicos, 1989). Equally, postmodernists have tended to reject what they regard as Marxism's blinkered focus on the economic production of value and want to avoid, at all costs, being associated with what Lyotard terms the "grand narratives" of modernity (Lyotard, 1984). These narratives, which would include both the natural sciences and Marxism, claim to provide universal or objective theories that transcend their social origins. By contrast, Michael Lynch claims that "[c]onstructivists openly espouse multivocality and anti-foundationalism; they defend subjugated knowledges against unitary notions of progress. In a word, they are incredulous toward master narratives" (Lynch, 1998: 13).

While Sessions classifies Marxists and postmodernists as extreme social constructivists, as a Marxist and materialist, Dickens happily targets both deep ecologists and postmodernists (strong constructivists) claiming that both positions entail a form of idealism. Postmodernists are idealists in the sense that they supposedly hold that nature is an infinitely plastic creation of cultural discourses. Deep ecologists are idealists in a different sense, they are unwitting dupes, entirely unaware that their idealized and romantic view of nature is actually an artifact of current social circumstances (Smith, 1997).

Such disagreements about the basic tenets of the debate indicate that, in reality, as Eder recognizes, these extreme "culturalist" and "naturalist" positions belie the real complexity of the issue since most theoretical paradigms incorporate elements of both. Despite the intense passions that have arisen in recent debates, attempts to draw strict dividing lines between positions are fraught with difficulty and often obscure real commonalties between them.

Some of the difficulties have arisen because of the tendency of both sides to elide the difference between constructivism as an epistemological thesis, of whether our knowledge of nature is socially constructed, and constructivism as an ontological thesis, of whether nature is nothing more than a social construct. If we choose to emphasize the import of social systems and practices on our phenomenal apprehension of nature, then we seem to be making an epistemological claim. But if we go further and emphasize the absolute inaccessibility of the noumenal, of a socially unmediated nature then this easily slips into an apparently ontological claim. This elision can work to both sides" polemical advantage. It seems to underlie Myeson and Rydin's (1996: 119) constructivist claim that "[e]nvironmental arguments *are* cultural arguments: culture is the subject and it is also the context." For them the environment *is* a "new rhetoric." Alternatively Dickens castigates Keith Tester for making what he takes to be ontological claims about the social construction of nature when Tester states that a "fish is only a fish if it is socially classified as one. . . . Animals are indeed a blank paper which can be inscribed with any message, and symbolic meaning that the society wishes" (Tester in Dickens, 1996: 72). From Dickens' Marxist perspective, Tester's "strong constructivism," would obviously count as a form of idealism in opposition to his own materialist ontology. There are, he says, "real differences between how people *construe* fishes . . . [and] how a fish is *physically constructed*" (Dickens, 1996: 73). On this issue Dickens is in absolute agreement with both Soulé and Sessions and they are surely right that the ontology of the world is not directly affected by the way society "chooses" to describe it.[11] A fish remains a fish, its physical characteristics remain unaltered, despite the possibility of its being construed in an almost infinite number of ways within different social formations.

Initially at least there seems to be radical differences over questions of ontology between strong constructivism and naturalism. But are all strong constructivists really making the ontological claims that Dickens takes them to be? In summing up the constructivist position William Cronon makes plain that he believes the "nonhuman world is real and autonomous" and that the nature that society reinvents refers to our conceptions of nature and not the nonhuman world's actual ontology (Cronon, 1994: 458). To concentrate on the manner in which nature is represented, spoken of, and utilized by social groups is to emphasize the import of the manner of our apprehension of nature on our activities, that is, on the way we do or do not value it. But it does not, or at least need not, deny nature's existence outside of discourse. Nor, despite Tester's bravado, does it entail that "anything goes."[12] Consider for example, the following: in some parts of South America, the Capybara, the world's largest rodent, is classified because of its aquatic habits as a fish. It is not incidental that this also means that, in a largely Catholic society, Capybara meat can therefore be eaten on Fridays. Of course, as a large furry mammal, the Capybara looks nothing like any other fish, but to concentrate on its ontology, on the question of what it "really" is, is to miss the reason for its particular cultural value in these societies. No one claims that the Capybara becomes a fish by changing its physical make-up but, while its biological nature is obviously not immaterial to its cultural classification (and its edibility), the constructivist is obviously right to emphasize that what *matters* here is, to a large extent, its cultural rather than its biological taxonomy.

Strong constructivists, such as Tester, may sometimes get carried away by their own rhetoric, but, generally speaking, constructivism is rooted in sociological appropriations of the phenomenological tradition and post-Saussurian linguistics rather than outright idealism.[13] In other words the constructivist approach is generally to suspend, rather than make, claims about the world's ontology since these kinds of claims are, rightly or wrongly, regarded as culturally bound and hence ultimately undecidable.[14] For example, John Hannigan's recent text *Environmental Sociology: A Social Constructivist Perspective* subsumes questions about the ontological status of the environment under questions about how nature is represented in various discourses. Following the "social problems" approach of Spector and Kitsuse he brackets out questions about the objective status of the environment and instead examines the claims-making activities of different participants in the debate (Spector & Kitsuse, 1967).[15] As Hannigan himself admits "[f]rom this point of view, the process of claims-making is treated as more important than the task of assessing whether these claims are truly valid or not" (Hannigan, 1995: 32–33).

So when Sessions sees constructivism as guilty of an "ad hominem" fallacy, providing an "analysis of truth . . . in which the emphasis is not on *what* is claimed . . . but on *who* or which social group says it" (Sessions, 1995: 15), he is both right and wrong. He is right insofar as constructivists do indeed tend to emphasize the claims-making activity itself as a locus of study, but wrong insofar as constructivists like Hannigan are not professing to tell us the truth of the matter at all. He is not making ontological claims. Sessions mistakes constructivism's methodological (and epistemological) presuppositions, which regard our knowledge of reality as inherently culturally bound, with an ontological position that claims that nature-in-general is nothing more than a social category, that is, that the truth of nature lies in what people decide to say about it. This difference may seem like a minor quibble but in fact it has massive implications for the relationship between constructivism and deep ecology.

It should by now be obvious that, in order to discuss these issues appropriately, we need a more sensitive typology of social constructivism, one that does not simply lump together Marxist, postmodern, and other approaches. The fact that these positions all problematize deep ecology's claims to understand nature in itself is not sufficient reason to regard them as equivalent, since they do so on different grounds. At the very least, I would suggest that we need to distinguish between those who are making ontological claims and those who espouse a methodology that regards discourse about the social/natural world as a closed system that can be approached without reference to the claims it may make about this underlying ontology.

Even within strong social constructivism, approaches that ignore the validity of claims are regarded as somewhat extreme. For example, Joel Best makes a distinction between "contextual constructivism," which recognizes the import of objective social conditions in evaluating claims about social problems, and "strict constructivists," who "are not interested in assessing or judging the truth, accuracy, credibility, or reasonableness of what members say and do" (Schneider & Kitsuse in Burningham & Cooper, 1999: 304).[16] I therefore suggest that we provisionally accept a typology that recognizes four positions ranging from Sessions and Soulé's "scientific naturalism," through "materialist" approaches like that of Dickens, through Best's "contextual constructivism" to the "strict constructivism" espoused (but not consistently practiced) by Tester, Hannigan, and others. Many of the criticisms directed at social constructivism by authors like Sessions and Soulé are really directed at only strict constructivism.

Deep Ecology's Knee-jerk Response to Social Constructivism

Even if we accept that the real debate between constructivism and deep ecology is not straightforwardly about the world's ontology, constructivsm still poses problems for deep ecology. Insofar as all varieties of social constructivism insist upon the constant presence of history, traditions, and power relations in all representations of nature, they expose something of the contexts in which discourses of nature are produced. The problem for deep ecology is that contextual, strict, and materialist constructivisms seem to reject *all* claims to grasp nature in a pure or unadulterated form.

Given this situation a rather strange alliance seems to be growing between certain deep ecologists and natural scientists seeking to assert the accuracy of their own ontological, epistemological, and ethical claims about nature. Some, though by no means all, deep ecologists seem unwilling to admit the cultural contextuality of their own discourse. They place themselves under the banner of nature pure and simple and on the side of a naturalistic realism, against its philosophical enemy constructivism. By doing so they hope to avoid the (cultural) relativization of values that they see as inevitably following in social constructivism's wake. But, I shall argue that deep ecology, if it is to be regarded as a political force for the future, must transcend this debate and avoid the short-term advantages some perceive to be gained from aligning itself with discourses of scientistic naturalism.

In their determination to defend nature's intrinsic value, deep ecologists like George Sessions and Paul Shephard are actively exploring the links between their own position and those natural scientists who have already been a focus for social and cultural criticism, for example, sociobiologists. Thus Sessions is happy to declare that genetics is central in defining the terms of any debate about social constructivism—"[t]he idea that humans have a genetically-based human nature is a bone of contention between deep ecologists and Marxist/postmodernists. The latter, harkening back to older Enlightenment/social science view's of humanity's uniqueness . . . hold that humans are not genetically hardwired for anything in particular" (Sessions, 1995: 15). He enthusiastically refers to Paul Shephard's work which suggests that "humans are genetically programmed for wild environments, and that . . . modern urban humans who have not bonded with wild nature are ontogentically stuck, remaining in some ways in an adolescent stage of human development" (Sessions, 1995: 15). This potential alliance is exemplified by Shephard's contribution to Kellert and Wilson's (1993) anthology, *The Biophilia Hypothesis*.[17]

Of course, the alliance with (natural) science is not limited to genetics. In a somewhat bizarre passage, Sessions castigates postmoderns (and presumably Marxists by default) as "urban theorists [who] have yet to deconstruct the biases of their own Enlightenment anthropocentric humanism, their antipathy to such modern sciences as anthropology, biology, genetics and structural linguistics which cast a new and important light on the universality of human nature and, perhaps most importantly, their own profound alienation from wild nature" (Sessions, 1995: 15). Apart from the fact that Sessions is seemingly unaware of the potential irony in what could be taken as a socially relativistic postulation that the "urban" origins of social theorists might contribute to their theoretical perspectives, he makes it clear that he regards his position as rooted in and validated by the privileged discourses of such sciences. It is presumably such lines of argument that have led critics of deep ecology like Ross to label it as "misanthropic," entailing a new "biologism"—an ecological version of all that was worst in the crude reductionist and determinist sociobiological analyses of social problems (Ross, 1994).

There are a number of problems with deep ecology taking this scientistic route. First it relies upon a homogeneous picture of scientific discourses, ignoring the very real debates *within* sciences. To do otherwise would of course be to admit that nature is indeed contested. But sociobiology and the question of genetic programming have been at the heart of some of the most controversial debates within science. Many geneticists explicitly deny that genetics can tell us anything much about the universality of human nature and tie such biological determinism to the rise of the new right in politics, racial stereotyping, IQ testing, and so on (Rose, Lewontin, & Kamin, 1985). What is at stake here is not simply a question of scientific fact, it is a case of real disagreement about what science says. To rely then on a particular reading of genetics—or any other science—is not to access a realm of objective knowledge but to be party to debates that have their own internal dialectics and external ramifications.

Second, it assumes that current scientific opinion gives us privileged access to the truth and hence provides a permanent anchor for our values. That is, it reifies current scientific theory as an accurate and unchanging representation of the world as it really is. Deep ecologists frequently cite ecological theory in debates over environmental evaluation. For example, Freya Mathews claims that ecologists believe ecosystems to be self-regulating entities and would confirm "the idea that ecological succession [. . . involves] a self-organizing strategy on the part of the ecosystem" (Mathews, 1991: 130).[18] Hence on her grounds ecosystems qualify as—"fuzzy—individuals with specific needs and specific ways of satisfying them" (Mathews, 1991: 132) and as such have intrinsic value. The ethical value of biodiversity can then be defended on the grounds of a postulated ecological relationship between ecosystem stability and species diversity. She constantly draws ethical conclusions from her use of ecology, claiming, for example, that "species loyalty—a limited prejudice in favor of one's own species in unavoidable, [. . . and] is ethically proper for a species whose population is ecologically viable" (Mathews, 1991: 129).[19] My point here is not that it is wrong to draw ethical conclusions from matters of fact (though it is certainly not straightforward), nor that the ethicists concerned have got their "facts" wrong (though deep ecologists frequently seem to be almost entirely ignorant of important debates in recent ecological literature[20]) but to emphasize that science changes, its opinions are not permanent. If deep ecologists really insist on tying their valuation of nature to current scientific views, then should one presume that their current values will be "hashed" if ecological theory alters? Will Walden Pond or Yosemite cease to have value if ecology deems that ecosystems cannot be interpreted as even "fuzzy" individuals?

Third, it misconstrues the real complexities of the relationship between science and deep ecology, making the latter subject to the very scientific rationality that, I have argued, many environmentalists feel is at least partially to blame for our current ecological predicament. There should be something terribly ironic in deep ecology looking for support in genetics at the very moment when massive sums of money are diverted into the human genome project and a whole new range of environmental and ethical dilemmas are posed by the release of genetically modified organisms into the environment. Certain kinds of genetics are prime examples of a Promethean and humanistic search for a technological fix to social and environmental problems. Indeed genetics raises the interesting specter of a technocratic solution to the problem Shephard raises about our lacking a necessary contact with nature, for example, by the provision of "gene therapy" to remove any such disposition.

Uncritical acceptance of science is a dangerous tactic, which could easily rebound on deep ecology, and calling upon the tarnished authority of science is surely ironic if not actually inconsistent with deep ecology's analysis of our current predicament. The (re)investiture of science with the crown of objective truth disregards the eloquent testimony to its fallibility provided by environmental disaster after environmental disaster.[21] It also gives scant regard to the many varied criticisms of scientific objectivity that fuel a whole academic industry in the philosophy of science. In other words it forces deep ecology to defend what may well be both environmentally and epistemologically indefensible.

Fourth, in complete contradiction to the stated aims of deep ecology, it unreflexively reinstates a humanist divide between nature and culture by emphasizing a dichotomy between the natural and social sciences. For example, Soulé's contribution to his edited collection is entitled "the social siege of nature" (Soulé & Lease, 1995). Both he and Sessions argue that these social critics are "humanists" and hence inevitably anthropocentric emphasizing "humanity's uniqueness [and] separation from nature" (Soulé & Lease, 1995; Sessions, 1995: 15). However, their own positions could equally well be labeled humanist since by anyone's definition, the dominant strands of Enlightenment humanism would have to be characterized by (a rather naive) faith in natural science as a human endeavor capable of revealing the truth about nature (Davies, 1997).

Fifth, it plays into the hands of those with an instrumental view of nature who are only too happy that deep ecology should paint itself into an unholy alliance with the often reactionary politics of sociobiologists and other biological determinists. By taking refuge in biological determinism in order to defend themselves against the supposed dangers of cultural relativism Sessions and colleagues are in danger of making deep ecologists everybody's favorite misanthropes (Bradford, 1989). But deep ecology is, and should be, a philosophy that informs radical (and not reactionary) social movements seeking profound changes in our relationships with nature—changes that also involve alterations in values, beliefs, politics, and economics.[22]

Sixth, deep ecologists have frequently looked to other premodern cultures, or non-Western religious traditions like Buddhism as exemplars of how we might become more attuned to nature. Thus they do accept that views of nature differ between and across cultures. But how are these differences to be explained, and how can we argue that our cultural perspectives are damaging, if, at the same time, we suggest that culture is not of prime import in constructing our values?

Finally, it ignores the very real diversity within constuctivist views, creating theoretical enemies where alliances could be made. The scientistic naturalism supported by some deep ecologists is unnecessary because not all forms of social constructivism are incompatible with the idea that nature is noninstrumentally valuable. I now take up this issue via a reinterpretation of the tensions between constructivism and naturalism.

Mediation, "Nature," and Value

I have argued that the spectrum of opinions about social constructivism reflects differing emphases on the relationship between nature and culture. Deep ecologists like Sessions have reacted to the relativizing effects of constructivism by trying to fix value in the permanence of nature's ontology. By linking ethical values to claims about this ontology they hope to defend nature's immutable value in and of itself. But doing so ignores the validity of the constructivist's cultural and political critique, a lesson that deep ecology must learn since as a radical philosophy dedicated to exposing the arbitrary nature of our contemporary cultural concerns, deep ecology has much in common with radical forms of social constructivism.

Constructivism has, in one form or another, played a vital role in the radical critique of oppressive social structures. Those in power always justify oppression, inequality, and injustice as the necessary outcomes of universal, objective, and unchangeable natural laws. Once this supposedly natural order is revealed as a social construct that is often arbitrary, always the result of special circumstances and everywhere serving the particular interests of the ruling elite, the possibility of change arises. When the ideological elements of previously held certainties are unmasked we can see the potential for making the social world anew and according to different principles.

Certain kinds of constructivism and deep ecology can both be regarded as "critical theories" (Vogel, 1996)—critiques of the current order of things, though both sides seem loath to admit their commonalties. However, the retreat of deep ecology into a reactionary scientism and the failure of Marxism and strong constructivists to take the noninstrumental values of nature seriously both indicate a lack of critical awareness on their own parts about their entangledness in social and natural circumstances respectively.

I want to tentatively suggest that we look at the debate over constructivism in rather different terms, terms that I will take up again in later chapters. Rather than stressing ontology or epistemology the argument might be seen to be one of location, of where values are produced and their degree of attachment to that locus of production, that is, their transitivity. In other words, the important differences between positions on the constructivist/naturalist spectrum might be seen in terms of the *medium* and *mediation* of value. To appropriate McLuhan's phrase for other purposes, in each case and for all parties concerned, the "medium is the message."

Three of the major players I have identified on the naturalist/constructivist spectrum—strong/strict constructivist's (including some postmodernists), Marxists, and scientific naturalists (including some deep ecologists) might be seen as locating the production and residence of values within their own favored fields, namely culture, the economy and nature respectively. And in each case values are not only products of either culture, the economy or nature but must be understood as saying something (carrying a message) about these fields. The actual disagreement between these parties then has little to do with ontology—they all recognize the existence of a material nonhuman world—but rather concerns where values are produced and what they express.

For strong constructivists values are constituted within the symbolic, ideological, and political order of society, they are therefore not only cultural products, their message is a cultural message. What is more, in cultures undergoing continual change, like our own, values too must inevitably shift. They cannot stay tied to the location of their production, they no longer represent fixed relations but must float free. Values are transitory and highly mobile shifting patterns of social mores, tradable fluxes. At the extreme they are, in Baudrillard's terms *simulacra*, constantly reproduced and changed without any necessary reference to an underlying "reality."[23] Nature here appears only as a sign with shifting patterns of meaning determined only by its position in its systemic relations to other signs.

Insofar as Marxism is concerned the ability to transform nature is (as chapter 3 illustrated) largely limited to the economic activities that form the base of society—where labor utilizes nature as raw material relying on its "ontological regularities" in order to harness its productive potential. The cultural superstructure's relation to nature-in-itself is always mediated via the economic base. Values are always, in the last instance, a product of underlying economic relations and can only affect transformations in nature via that economy (and vice-versa). The medium and the message for Marxism might be said to be those "relations of production" wherein labor produces *use-values*. For this reason, as Eder points out, Marxism is wedded to an instrumental evaluation of nature (Eder, 1996). From this perspective, as chapter 3 showed, the culturalist approach of strong constructivists privileges a system of exchange values that has lost its direct relation to reality, that is, the ontology of labor's productive activities.

Deep ecology sees nature as the ultimate source and location of values, including of course human nature. Nature is itself regarded as an autopoeitic productive field. In Rolston's words a "tree grows, reproduces, repairs its wounds and resists death. A life is defended for what it is in itself, without necessary further contributory reference. Every organism has a *good-of-its-kind*" (Rolston, 1997, 61). Once again the medium of values production is the message; nature is what carries and determines the allocation of value.

Perhaps viewed this way, in terms of the production and mediation of values, we can at last begin to see the wood for the trees. Previously the debate was envisaged as an ontological and epistemological conflict between naturalists who (claim to) have an understanding of the way nature really is and can hence recognize its true intrinsic value and constructivists who claim nature, and the way we value it, are simply manufactured by human societies. The issues may now seem both more complex and more amenable. More complex because each perspective is now seen to be indicative of differing analyses of social/natural relations. More amenable since we no longer need to rely on driving spurious ontological wedges between positions but can recognize that each position is, in a sense, a critique of the naturalness of the current status quo.

All sides have to recognize something of value in each other's analyses. There is surely something in postmodern claims that values are indeed becoming less and less tied in place, their transitory nature is oft remarked in all fields and agreed by those right across the naturalist/constructivist spectrum. To even account for such a situation naturalists and Marxists have to recognize the role of culture in this ethical fragmentation.[24] Similarly deep ecologists and constructivists are able to recognize the powerful role played by our current economic system in bringing about changes in our relation to nature. Why then should Marxists and strict constructivists not recognize that nature too produces changes in culture and the economy?

I have suggested that even strict constructivists rarely if ever hold the view that society can ontologically reconstruct nature by a mere act of definition. In this sense there are no fundamental ontological differences between the positions, as is sometimes claimed to be the case. We are not dealing with a debate between realists and idealists. However, constructivism does (rightly) contend that we have no direct access, even via the natural sciences, to an unmediated nature; that deep ecologists, intent on highlighting the reality of nature, cannot simply ignore the reality of cultural or economical differences. The reverse side of this coin is that, insofar as constructivists do admit the ontological reality of nature as well as culture, they cannot justifiably chose to simply bracket it out—to ignore nature's productivity. To do so would indeed be idealism.

Similarly, we may have no direct epistemological access to nature but it is equally the case that we have no unmediated access to the reality of society, of culture, or the economy either. We may not be able to grasp nature except through the mediation of society but nor can we claim to understand society without taking into account the medium of nature. Once we recognize that there is no ontological dispute here, that both nature *and* society are *for real*, then epistemological debate is also seen to work both ways. Contra the naturalists, there is no master discourse that can assure agreement about nature. Contra the constructivists culture is not the only medium of our existence and the only thing that is constitutive of discourse.

And here we return to the problem of the dialectic. Nature is indeed contested, but it is also a contestant, a constitutive part of the medium of our existence. As such, there is nothing nonsensical in valuing those parts of nature, which we choose, or are brought, to recognize either through our own, or nature's activities. Insofar as they ever could be, these values are biocentric, that is, genuinely concerned to recognize the relative autonomy of trees, and so on, as being just as much entities in there own right as are humans.[25] These values may not be intrinsic in the absolutist and universal way that certain ontological forms of moral realism may suppose, they are not eternally fixed to some internal essence of the thing concerned, but nor are they just the products of human whim, will, or work. Nature too participates in this dialectic.

There is nothing to stop a genuinely contextual constuctivism from recognizing this point. Indeed, to do otherwise is to reduce the values we recognize in nature to either transitory eddies in culture's increasing self-referential circulation or to instrumentalize them. It is to divorce society from the natural part of the medium of our existence, thus reinforcing life in the dream world of hyperreality or the nightmare of economic reductionism. But finding our place in, and as part of, nature requires that we recognize the value of more than the products of our own limited agency, that we stop seeing our own activities as the be-all and end-all of the world. Deep ecologists rightly insist that, despite the metaphoric layers that condense around and colonize it for particular social purposes, nature is never simply a social product. It is neither a *tabula rasa*, a raw material wanting the in(ter)vention of labor to inscribe it and hence interpellate it into the world, nor is it, what is little better, a text passively awaiting the imaginative inventiveness of the human critic.

The recent awakening of the social sciences to "the environment" as a field of inquiry and source of political activity has predictably led to attempts to squeeze the question of nature into social theory's own prescribed and explicitly anthropocentric problematics. It is perhaps particularly ironic that those strict constuctivists, the cultural critics who, after arduous struggles freed themselves (to a greater or lesser extent) from the economistic reductionism of the social sphere, are overzealous in their attempts to reduce natural to cultural forms—to lose nature's profuse fecundity in the tangles of often naive historicisms. On the other hand, so long as deep ecology remains attentive to its own debt to society, its eloquence on behalf of nature need no longer imply a kind of silence about injustice.

In conclusion, we *must* make the (ethical) attempt to speak of trees and speak of them care-fully as having what we might term "constitutive values" rather than intrinsic value. These constitutive values are not reducible to either use or exchange values—but nor do they stamp nature with that timeless and universal guarantee of worth that some deep ecologists vainly demand. Given the right social and natural circumstances such values can be found by those who seek them, so long as we recall that they are sensitive products of circumstance. Nature's value does not reside within trees, waterfalls, badgers, or bats but are constitutive of the ethical attempt to recognize such things for what and who they are. Nature calls on us to respond appropriately. Ethics requires attentiveness, a relation of appreciation rather than appropriation; this ethical response is the topic of subsequent chapters.

5

Environmental Antinomianism: The Moral World Turned Upside Down

If, as previous chapters have suggested, contemporary theoretical problematics are largely incapable of understanding, let alone assisting, radical ecology, this can only be because such environmentalism sets out to undercut our current ways of thinking and our current "forms of life." The alternative worldviews of radical ecologists are made manifest at all levels; in the personal and practical politics of nonviolent direct action, the countercultural interventions in literature, poetry, and electronic media, the development of alternative economic systems, as well as in the conceptual and theoretical reorientations that arise from questioning where we are, how we came to be here, and what is to be done.

This questioning goes to the very foundations of our thinking and its implications are nothing less than revolutionary. Up until now, I have concentrated on mapping our current conceptual topography and illustrating the pervasive influences of anthropocentrism and rationalization in various modes of theoretical endeavor. To this end I have sought to deconstruct a number of different discourses in ethics, politics, and social theory that have marginalized nature and confined our conceptions of ethics. But now I want to begin to expand upon the tentative suggestions I have made as to how an environmental ethos might be reconceptualized in terms of the production of ethically charged spaces.

I hope to do this in two stages, both of which are inspired and informed by recent environmental protests, including antiroads protests in Britain. First, (in this chapter) I intend to show just why radical environmentalism might be so disruptive to the current social and theoretical order of things. Put simply, radical environmentalism is anarchic, that is, it sets itself against all attempts to manage, control, and order our relations to nature—whether these are associated with hierarchical and authoritative institutional structures or modernist rationales. Second, I try (in chapters 6, 7, and 8) to articulate a radical environmental ethos in terms of an "ethics of (particular) place(s)," contrasting this with the linear, abstract, context free space that modernity presumes and produces. This analysis seeks to reconstitute ethics environmentally in order to counter the current enclosure of the moral field within economistic and legal-bureaucratic frameworks.

Those nomothetic (lawlike) models that represent environmentalists as (a) seeking to extend current legal/bureaucratic frameworks to nature, or (b) drawing moral conclusions from natural laws have misunderstood its intent due to the limits of their own thinking. I hope to show that radical environmentalism actually has many parallels with past antinomian protests in rejecting the supposed ethical authority of those social institutions and theoretical frameworks that attempt to define and impose norms of belief and behavior. Radical ecology "places moral commitment above positive law"[1] and is characterized by a "hermeneutics of suspicion" directed toward the establishment in all its forms and extending to all attempts to "lay down the law."

Antinomianism

> *Albion goes to eternal death . . .*
> *No individual can keep these Laws, for they are death*
> *To every energy of man and forbid the springs of life*
> —William Blake, *Jerusalem*

> *True freedom lies in the enjoyment of the Earth.*
> —Gerald Winstanley in Hill, *The World Turned Upside Down*

Many people, at one time or another, find their personal values clashing with those that find expression in and are enforced by the Law. More often than not this conflict results in nothing more than a grudging acceptance of life's inherent injustice. Sometimes, if values are held dearly enough, that injustice can evoke resistance from the individuals concerned. This may take the form of a principled refusal to comply, for example, withholding certain taxes, or a more confrontational strategy as demonstrated in the recent street protests in Seattle against the World Trade Organization. However, the espousal of avowedly antinomian sentiments, literally being against the law per se, is far less common; indeed for many it is unthinkable. After all, the law, whether codified in religious texts or in the secular law of the land, is seen to provide a stable normative framework, a set of explicit guidelines regarded as necessary to ensure society's continued functioning. The Law, quite literally, defines the boundaries of our existence, both in terms of specifying the precise nature of proscribed activities and in prescribing society's moral horizons. Those forces of law and order that claim to embody and uphold the law's principles also assume the right to police and to pass judgment on others. They ensure that those who ignore or flout the law's authority, and thereby upset the regularity and regulation of society, are brought to justice and suitably punished.

Yet the very requirement for a body of law points to the existence of contesting forces, to a constant undercurrent that would disrupt society's apparently tranquil surface. The greater the divergence of interests bound together, the more contested the form of society, the stronger the law must be in order to dam the otherwise free-flowing passions that might be released. Unsurprisingly, those who find their interests well represented and protected by the law tend to emphasize the universality of its principles and applications. Those with divergent interests will, on occasion at least, find it necessary to speak of its hegemony in more qualified tones. To some this questioning of authority applies only to specific laws seen as incompatible with a social ethos with which they generally concur. Others, more at variance with the dominant ethos, may question the legitimacy and form of the entire body of the law, comparing it to its detriment with some higher and more perfect order whether worldly or unworldly. Yet antinomianism goes a stage further. It not only questions the legitimacy of a particular law or set of laws but the idea of "legitimacy" itself. It brings into question the very idea of policing boundaries and limiting horizons.

The outlines of these positions are exemplified in Foucault's "On Popular Justice: A Debate with Moaists." Far from supporting a liberal tinkering with particular laws, the Moaists argue against the legal system prevailing in capitalism, contrasting it with the idea of a people's court. They believe that "it is necessary that there be some legal authority so that the diverse acts of [postrevolutionary] vengeance should be *in conformity with law*, with a people's law which is now something entirely different from the old system of [Chinese] feudal law" (Foucault, 1980: 2–3). By contrast Foucault argues that popular justice is and should be "profoundly anti-juridical and is contrary to the very form of the court" (Foucault, 1980: 6). Foucault rejects the idea of a "*regulatory instance*" preferring to speak of an "*instance of collective political elucidation*" (Foucault, 1980: 29). While being intentionally vague about precisely what form this elucidation might take, he states "[a]bove all it is essential that the stick be broken" (Foucault, 1980: 32). The Maoist idea of a "proletarian counter-justice is a contradiction; there can be no such thing" (Foucault, 1980: 34).

Here Foucault in effect espouses an explicit antinomianism, the rejection of the idea of all overarching systems of law (at least insofar as they are embodied in juridical models).[2] He regards it as inevitable that if justice takes certain authoritative forms then it will simply reiterate and replace one set of constraints on the free expression of the vox populi with another along similar lines.[3]

Such antinomianism has taken different forms in different times and places. Certain early Gnostic sects provide a good example of the rejection of the authority of the moral law even when it apparently has the textual backing of scripture. For example, the Ophites glorified the serpent that tempted Eve to break God's commandment not to eat of the Tree of Knowledge as the liberator of humanity. Viewed from outside such antinomianism is not only heretical but also doubly dangerous since it seems a small step from advocating *amorality* to acting *immorally*. The rejection of the moral law seemingly suggests that anything is permitted and antinomianism thus often finds itself associated, whether justifiably or unjustifiably, with libertinism. For example, the Gnostic Nicolaitans (whether or not they actually existed) were accused of advocating a "return to pagan morals especially in matters of sexual practices" (Cross, 1966: 958). Gnostics believed in a complete divide between the divine substance (including the soul) and the material world (including the body), which they regarded as a creation of evil. Irenaeus could therefore accuse the Valentinians (another Gnostic sect) of depravity since, as they believed that "the spiritual element cannot be corrupted whatever [earthly business] it may be involved in," they were seemingly "allowed to do everything, to violate every norm for they are above every ethical and legal convention" (Irenaeus in Filoramo, 1991: 186). In particular Irenaeus claimed that the Valentinians eat flesh dedicated to idols, corrupted women, and even went to pagan events such as the circus and the theater!

In the fourth century Epiphanius, the future bishop of Salamis, claimed to have come across a hedonistic Gnostic sect who "have their women in common [. . . and] serve up lavish helpings of wine and meat *even if they are poor*. When they have had their drink and filled their veins, as it were, to bursting point, they give themselves over to passion [. . . and] indulge in promiscuous intercourse." Though it "truly shames" him, Epiphanius does not recoil from reporting their activities "so as to arouse in [. . . his] readers a shuddering horror of their scandalous behaviour." A description follows that details fornication that "they deny [. . .] is for procreation. They practice the shameful act not to beget children, but for mere pleasure." They are also accused of the ritualized eating of sperm and menstrual blood, contraceptive practices, abortion, and even consuming the aborted embryo garnished with perfumed oils, honey, pepper, and other spices. All this they "consider the perfect Passover" (Epiphanius in Filoramo, 1991: 186).

Whatever the truth in such accounts presented by their institutional adversaries, there is no doubt that the Gnostic's antinomianism could and more frequently did lead in entirely different directions, namely toward sexual abstinence and asceticism. However, critics of antinomianism always emphasize the "inevitable" slippage from *a preceptual amorality* to an *unlicensed and licentious immorality*. Something of this charge of libertinism re-emerges if we look at later cases of antinomianism. The so-called Ranters, nondenominational preachers closely associated with the revolutionary Leveller and Digger movements in seventeenth-century England, also found their personal sexual morality questioned. Some indubitably did preach free love. The Ranter Abiezer Coppe claimed to be able to "kiss and hug ladies, and love my neighbours wife as myself, without sin" (Coppe in Hill, 1996: 218).[4] More often though such accusations were either attributed to Ranters and revolutionaries by their opponents, exemplified in the invented song "Sisters you may freely do't" (Hill, 1996: 218) or tell us more about the preoccupations of their accusers then the antinomians themselves. In 1746 John Wesley pressed an antinomian as to whether his refusal to recognize any moral authority meant that he had "a right to all the women in the world?" As Christopher Hill reports, the antinomian tactfully replied "Yes, *if they consent*" (Hill, 1996: 224). It should be no surprise that the perceived refusal to accept the normative regulation of sexual practices should be such a focus of moral censure. Sexuality is a frequent focus for moral panics and a key indicator of a threat to moral stability. (The recent, but by no means exceptional, media interest shown in the extramarital behavior of certain prominent politicians attests to this.) What is more interesting is the common association in the minds of all opponents of antinomianism that the rejection of the moral law, the refusal to seek permission from authority for one's activities and beliefs necessarily entails permissiveness. People, it seems cannot be trusted to control themselves but need to be supervised to make sure that they do not stray from the straight and narrow.

The period of the English Civil War saw a veritable plague of antinomianism, both political and religious. The authority of the church and of the king was challenged by radicals in terms of a rejection not of particular power structures but of all authority. After the regicide, the attempt by Parliament to impose its own jurisdiction on the *unruly* Levellers, Diggers (or True Levellers as they were sometimes known), and Ranters inside and outside the New Model Army met much resistance. For as Sabine (1963: 488) remarks, "[f]rom the Levellers' point of view there was no more merit in Parliament's claim to sovereign power than the king's."

The languages of political and religious antinomianism were inextricably linked. For Gerrard Winstanley, the Diggers' chief pamphleteer, Jesus Christ was the head Leveller. The Diggers' conception of Christ and Christianity was far from that of the *established* church. For example, as Christopher Hill describes it, "[o]ne Sunday in March or April 1649 the congregation of the parish church of Walton-on-Thames was startled to see the church invaded by a group of six soldiers. . . . The soldiers . . . announced that the Sabbath, tithes, ministers, magistrates and the Bible were all abolished. On Sunday the 13th April—quite possibly the same Sunday—a group of poor men . . . collected on St. Georges Hill in the same parish and began to dig the waste land there" (Hill, 1991: 110). It was probably not, as Hill notes, accidental that they chose to begin work on a Sunday. The occupation and tilling of St. George's Hill was a symbolic and practical gesture intended to mark a radical opposition to the private appropriation and enclosure of common lands and a direct affront to the authority of church and state made in the cause of radical freedom. In Winstanley's words "[f]reedom is the man that will turn the world upside down, therefore no wonder that he hath enemies. . . . True Freedom lies in the community in spirit and community in the earthly treasury, and this is Christ the true man-child spread abroad in the creation" (Winstanley in Hill, 1991: 107).

Against the authority of the law the antinomian radicals preached free grace, that Christ's death had extirpated all sins. Christ was in everyone and belief was all that was necessary to assure salvation. The Ranter John Saltmarsh preached that "[t]he Spirit of Christ sets a *beleever* as free from *Hell*, the *Law*, and *bondage*, as if he were in *Heaven*, nor wants he anything to make him *so*, but to make him *believe* that he is so" (Saltmarsh in Thompson, 1994: 15). This antinomian spirit is echoed in Milton's *Paradise Lost*, where the archangel Michael says of Christ

> To the Cross he nails thy enemies,
> The Law that is against thee, and the sins
> Of all mankind with him there crucified (xii. 415–19)

Antinomianism sought to replace the doctrine of election, of God's choosing some in preference to others, with the simple and egalitarian need for belief. Indeed many held, as the passage from Milton suggests, that Christ's sacrifice atoned for all sins, even those of unbelievers. Nor, as E. P. Thompson points out, was the removal of the moral law seen to leave a vacuum. Rather it was to be replaced with love, a love that knew no bounds, a love from the heart, an enduring sympathy with one's fellows.

Repression was inevitable since antinomianism by its very nature is, as Christopher Hill argues, "a heady doctrine, *philosophical anarchism* which could call into question all earthly authority, all earthly law. In the "wrong" hands it could be profoundly and intolerably subversive of law and order" (Hill, 1996: 220). However, Thompson has shown how this antinomianism, forced by the state's military and legal repression to go underground, did not just die but continued through the seventeenth and eighteenth centuries in sometimes obscure sects like the Muggletonians, who, Thompson argues, may have had a profound effect on the writing and art of radical figures like William Blake (Thompson, 1994).

Like their predecessors, the eighteenth-century antinomian descendants of the Ranter and True Leveller traditions included "reason" in their rejection of authority, that same reason that had been crowned by the Enlightenment philosophers as the new secular principle capable of ordering psyche, society, and even nature itself (Thompson, 1994: 112). These "serpent reasonings" were both the cause and effect of a base materialism. For example, William Blake rejected Francis Bacon's *Essays* as "good advice for Satan's Kingdom" (Thompson, 1994: 112). Blake felt only "contempt and abhorrence" for such thinkers since they "mock inspiration and vision" (Thompson, 1994: 113). Of course, given the increasing hegemony of rational discourse, such opposition risks being labeled as mere irrationalism, or even madness, as indeed was the case for Blake himself. But this counter-Enlightenment ethos makes sense given that, from a radical perspective, Newton, Hobbes, and Locke were all members of polite society and their reason could be viewed as an ideological principle that they utilized to reorder society in their own class image.

In some cases, like that of the Muggletonians, this rejection of reason extended to education in general. Education was associated with the Fall and so with both the loss of innocence and the materialism of those who focused entirely on this world, rather than on the utopian possibility of building a New Jerusalem. It was an instrument of social control. Thompson quotes some lines found on a Muggletonian Song Book.

> By edducation most have been mislead
> So they believe they were so bred
> The Priest continues what the Nurse began
> And thus the Child imposes on the Man (Thompson, 1994: 87)

This is, of course, a theme which finds echoes in many eighteenth-century thinkers, even those so closely tied to the Enlightenment as Rousseau, who saw *true* education's primary concern as maintaining a space for individual autonomy in the face of social pressures and evils.

This counter-Enlightenment rejection of reason is part and parcel of antinomianism's constant counterhegemonic strategy; the reaction against and subversion of those principles, whatsoever they might be, that seek to impose order from above or outside. It is vital that we understand this opposition to reason in its own terms and not in terms of reason's own ideology, which sets up an opposition between rationality and irrationality, and reason versus myth, superstition, and tradition. It is too easy for those schooled in an Enlightenment concept of reason to accept its propagandistic dismissal of its opponents and take its liberatory claims at face value. From a certain antinomian perspective reason is *profane*, it is of the head and not the heart, it is an evil inherently associated with the instrumentalization and desacralization of the world and deeply implicated in the working out of an emergent capitalism and of modernity in general.[5] Far form being a neutral tool, reason is an instrument, a tool of the devil. For the antinomian, reason was the secular ideological equivalent of Presbyterianism—a state religion whose primary aim is to reimpose order on the unruly and to channel and stifle dissent. However bizarre the discourse and imagery of seventeenth- and eighteenth-century antinomianism may seem to the modern ear and eye, it is in many respects more consistent than rationalism in its refusal to make a god out of reason, to imbue it with an unquestionable authority, as having a unique ability to divine the world's order and a divine right to order the world.

From this historical sketch we can see that the key constituents of antinomianism might include: the *rejection of political authority*, in particular as it is embodied in the law and its associated institutions; the *rejection of all moral authority* and a consequent refusal to accept the *imposition* of moral norms whether secular or religious; an emphasis on the *free individual* as responsible for creating their own space of engagement with the world; a deep-seated suspicion or even outright *rejection of rational authority* insofar as it claims to be the sole arbiter and administrator of our lives; and last, but not least, a utopian belief in a *New Jerusalem*, a new world *without an order* and without hierarchies. In short, of liberty and love against the law, of inspiration versus reason, and a vision of a world turned upside down.

Such antinomianism is indeed quite literally an-archy, an unfettered and unconditional rejection of all ruling principles in the name of freedom. (One is reminded of Michael Bakunin's statement that "liberty is the rebellion of every man [sic] against the tyranny of men" [Bakunin, 1973: 149]). Contra liberal or social democratic conceptions of law and morality, it refuses to recognize either "contracts" or "charters." It offers the individual the latitude to control their own destinies, to be responsible for their own moral conduct.

Of course antinomianism could, and often did, lead to a variety of political quietism, an otherworldliness, born out of despondency or the fear of repression. The Muggletonians retired to the backrooms of public houses to drink and sing their songs together, a closed society refusing to evangelize in any way except by keeping their texts in print, finally dying out in the twentieth century. Yet it should be remembered that in its radical heyday antinomianism was a powerful influence seeking to build the New Jerusalem here in Albion and not passively waiting for the afterlife. And, in differing periods of unrest, as in the Jacobinism following the French Revolution, it could and did burst back into flame.

Environmental Antinomianism

Given this characterization of antinomianism it is not hard to identify a strand of radical environmentalism that features many and often all of these elements, even if they often appear implicitly and side by side with more reformist agendas. (As the True Levellers did with the less radical Levellers.) For example, on December 20, 1994 protesters occupying part of Pollok Park, Glasgow, in the path of the proposed M77, announced the formation of the Pollock Free State.[6] The Declaration of Independence stated "at this moment in time, we believe the ecological holocaust facing our lands and wider environment to be so great, that it is our right, our duty, to *throw off* such forms of government that allow such evils to continue" (Seel, 1998: 109). The "rights" and "duties" here are self-generated and self-imposed, they explicitly reject any requirement for morality to be guaranteed by the external authority of a system of law. Indeed the point of the declaration is to symbolically express a rejection of the government's supposed moral and political authority. As Ben Seel, who lived in the camp writes, "[t]here was no faith whatsoever among the core group in "representative" channels. The participative empowerment they sought to encourage was extra-institutional" (Seel, 1998: 121). The community was intentionally nonhierarchical and food was paid for communally. Within the community money had little role because of its obvious associations with power and control and the predominantly financial motivations for environmental destruction. This rejection of any implied loss of communal autonomy is indicated by the fact that the camp turned down an offer of financial support and publicity from Greenpeace despite the offer having the backing of Colin, one of the camp's founders. There were no codified rules or proscriptions within the camp with the exception of a self-imposed ban on on-site alcohol consumption in order to discourage local alcoholics from seeing the camp as an alternative late night drinking venue. The philosophy of the camp was summed up by one participant, Anna, as "[d]o it yourself" (Seel, 1998: 121).[7]

These patterns are repeated in other environmental protests though with different emphases (Evans, 1998). Thus Merrick, a protester against the Newbury Bypass in southern England, states: "[l]aws are made by a few people to protect their own interests, and laws are only passed to benefit the wider society when those in power can no longer refuse. I have no problem with breaking the law if the law is wrong. . . . No what we are doing here is proving these laws wrong" (Merrick, 1997: 12).[8] At times Merrick's emphasis seems to be much more on the rejection of individual laws rather than the law itself. "We are part of the movement to reform unjust laws." However, like those in Pollok Park, he recognizes that the creative opposition to environmental destruction entails a rejection of current social conventions and orders, the building of an alternative culture, and this is often expressed in opposition to the law. "[I]t wasn't just the CJB [Criminal Justice Bill] anyway—that's the real strength of Newbury and all the 90's counterculture protests—we're not fighting one thing we don't like; we have a whole vision of how good life could and should be, and we're fighting anything that blocks it. This isn't just a campaign, or even a movement, it's a whole culture" (Merrick, 1997: 12).[9] This culture is not to be founded on any legalistic authority but on love, of fellow humans and nature. "We'll be fighting with the strongest weapon of all. We'll be fighting with love" (Merrick, 1997: 9).

A love of nature, is regarded as the source of both the creative inspiration behind radical environmentalism and that place which shelters and nurtures individual freedoms. It offers possibilities for widening horizons, escaping current social structures, and developing alternatives. This antinomian notion of individual liberty in nature is expressed differently in different settings. Dave Foreman, one of the founders of Earth First!, gives it a particular North American slant: "Wilderness is America. What can be more patriotic than *love* of the land? We will be Americans only as long as there is wilderness. *Wilderness is our true Bill of Rights*, the true *repository of our freedoms*, the true *home of liberty*" (Foreman in Lee, 1995: 43).[10]

Of course, not all supporters of environmentalism are happy with antinomian sentiments. For example, Bron Taylor, a Canadian academic interested in radical environmentalism, criticizes the views of many of the Earth First! activists he met who "view their actions as part of an anarchistic assault on the industrial state itself, believing that a non-coercive and non-hierarchical society is required if people and nature are to be reconciled" (Taylor, 1997: 30). Taylor argues for a commitment to the "very democratic process that many of them have abandoned as corrupt"

(Taylor, 1997: 34–35). However, the fact that he thinks it necessary to argue such a case is symptomatic of the widespread opposition to reformism among activists. After all the slogan of Earth First! is "*No Compromise* in Defence of Mother Earth!" Taylor is also happy to admit that "conscientious lawbreaking" is sometimes a political necessity.

Despite the fact that rationalism is much more pervasive and prevalent today than it was in the seventeenth and eighteenth centuries, radical environmentalism continues to bring reason's authority into question. This often takes the form of emphasizing *feeling*, not in order to reiterate a simplistic dichotomy between reason and feeling, but as an attempt to re-evaluate and recognize that which has been suppressed by reason's almost unchallenged hegemony. Many radical environmentalists and deep ecologists emphasize the need for individuals to feel involved with nature, to commune with it. And, if these kinds of expressions and relationships are considered abnormal by society's standards of rationality, then so much the worse for normality. The language used often employs a religious metaphorics without implying any religiosity. For example, Dave Foreman, like many, speaks of receiving signs from nature, of being inspired by a freak snowfall during a speech or the appearance of a grizzly bear and cubs at the roadside after being arrested for trying to preserve their habitat in Yellowstone Park. "Rationality be damned. The ecstatic pagans on that bus had just received a sign from the wild—*keep on*!" (Foreman in Lee, 1995: 94).

As with previous expressions of antinomianism, the actions and rhetoric of radical environmentalism are often regarded as signs of irrationality. Those who contest environmentally damaging developments are portrayed as abnormal. Normal people don't dig and live in tunnels called "Cakehole" or "Sir Cliff Richard OBE Vegan Revolution" as those protesting against the building of a second runway at Manchester airport did. Normal people don't occupy camps called "Radical Fluff," "Trollheim," or "Granny Ash"; nor are they called "Animal" or "Swampy," or "Hengist" like those at Newbury. Who in their right mind would proclaim themselves to be King Arthur Pendragon and give their address as Camelot? (Though the apparent irrationality of declaring oneself a mythical king at least gives one the self-appointed right to oppose the law. Upon being stopped at the security desk of the High Court for carrying a sword King Arthur declared himself naked without it and took off his clothes in the lobby. He was eventually allowed in, with his sword.) In Merrick's words, "All these laws treat us as if we should have the same aims, principles and actions. But they can't do that with Arthur. I wouldn't be his kind of individual but I admire his striking some witty blows for individuality" (Merrick, 1997: 117). Despite being seen even by his fellow protesters as a "bit of a loon" King Arthur gained respect by putting himself on the line. He "wasn't above digging the shit pit at Camelot Camp" (Merrick, 1997: 117).

This rejection of authority is again evident in the lyrics of musician and ex–The Teardrop Explodes frontman Julian Cope. His poem "Socrates Mine Enemy" expresses his opposition to both straight roads and "straight" thinking. Julian's support for the roads protest at Newbury included playing a benefit gig, providing food, and "weirding out the security guards" (Incidentally two of his albums are entitled "Jehovakill" and "Autogeddon.") His well-known appreciation of ancient stone circles also finds an echo in the "to Pollok with love" campaign when Earth First!ers organized old cars to be driven up from England via antiroad camps and then turned them upright, cemented into holes in the path of the M77 to form a carhenge that was "ceremoniously set on fire" (Seel, 1998: 114).

Whatever their apparent irrationality, such actions are inspired and inspirational. The normally temperate Hugo Young columnist for the British center-left broadsheet *The Guardian* could write that "[t]he most awe inspiring political figure I've met this year goes by the name of Balin. At a time [just prior to the British general election] when party leaders are hyping their sincerity machines into overdrive. . . . Balin makes frauds of them all. For 16 days and nights, he lived 10 feet above the ground, dangling from a scaffolding tripod . . . neither Mr. Major nor Mr. Blair has ever made a personal sacrifice for anything he believes in. Balin is 31 and an educated, cheerful man but his conduct looks like madness. . . . Although he looks like a crank . . . he is in fact the prophet of the coming orthodoxy [sic]" (Young in Merrick, 1997: 36).

The law's response to such protests also makes plain the connections with the seventeenth-century acts of enclosure that were the main focus of the Diggers antinomian opposition; a connection made consciously in environmentalists' reoccupation and cultivation of St. George's Hill and a deserted brewery site in Wandsworth, South London under the auspices of the "This Land is Ours Campaign."[11] At Manchester airport protesters against the building of the second runway found themselves the subject of a "runway exclusion zone." They were prosecuted under the Trades Union and Labour Relations (Consolidated) Act 1992; legislation originally designed to stop secondary picketing (Wainwright, 1997).[12] The sheriff's officers and a hired army of uniformed and masked assistants dispersed the camps much as Cromwell's troops had dispersed the Diggers at St. George's Hill, though marginally less violently (Ward, 1997).

Unsurprisingly, environmental outlaws are beyond the pale in other ways that have interesting parallels with past accounts of antinomians.

One British tabloid reported a "Secret Hippy Orgy Under Runway Two" although it found it difficult to decide whether those involved were a "band of randy eco-warriors" including one conveniently named "Pervy," or a "bunch of smelly protesters." The article was accompanied with a picture of a semiclothed woman holding a spade entitled "DD-igging in" (Sunday Sport, 1997). The rather more upmarket *Observer Review*, disdaining such nefarious interests, presented a hugging couple, Denise and Andy (Grandpappy), entitled "The Tunnel of Love." While the article is generally favorable and constrained by its broadsheet sensibility, it nonetheless plays with sexual innuendo—"For Denise and Grandpappy, the earth moved under Manchester Airport"—and emphasizes the "oddity" of the older and pregnant Denise coping with morning sickness in a tunnel and having a relationship with a younger man (Gerrard, 1992). As a further example, Charlotte Raven in her *Guardian* column reveals that, despite the fact that "[h]e shuns designer labels and his trademark single dreadlock is a matted mess . . . [Swampy, a protester famous for his tunneling exploits] has achieved so many female admirers that he's recently been touted as a sex symbol" (Raven, 1997).[13]

Despite these difficulties radical environmentalists maintain their utopian opposition to the forces of law and order even as they are ripped from their eyries and pulled from their caves. From his perch in a tree about to be felled, Merrick planned to play Julian Cope's "Greedhead Detector," alongside a collection of other apposite songs including two different versions of Jerusalem. In this way they maintain a vision that finds salvation in not building on England's, or anyone else's *green and pleasant lands*.

Moral and Political Dimensions

While it may be illuminating to simply draw parallels between the antinomianism of the seventeenth- and eighteenth-century radicals and today's environmentalists, this anamnesic exercise has wider import. Recognizing this genealogy may be vital if radical environmentalism is to retain its life-blood in the face of the many attempts to either repress it, dilute its message, or divert it down already existing roads. There are, I think, two main theoretical reasons for stressing this antinomian strand of radical environmentalism; namely to disentangle it from political and from moral misunderstandings and misrepresentations that insist on regarding it nomothetically as either (a) seeking to *rewrite natural laws* or (b) setting out to *reform social laws*.

The first strategy attempts to reduce radical environmentalism to a form of so-called ecologism—defined as replacing social laws with a willing submission to natural laws. One such example is found in the work of Tim Haywood. He sets up an opposition between managerial approaches to environmental problems, which are obviously rejected by radical environmentalists, and an ecologism characterized as a discourse of naturalistic necessity. Ecologism, he says "invokes a much stronger, "internal" relation between ecology and politics: on this view ecology does not simply furnish "issues" for politics to deal with it actually yields *imperatives*" (Haywood, 1994: 11). Haywood's target is a variety of environmentalism that would subsume ethics under "natural laws" as a series of political, legal, and moral prescriptions read off from a reified understanding of the world's natural order; a call for moral norms and governmental policies to be determined by such things as resource scarcity, carrying capacity, and so on. He characterizes those environmentalists who reject managerialism as "being more sanguine than there is good reason to be about the possibility of drawing clear and *unequivocal normative desiderata* from *ecological realities*" (Haywood, 1994: 11).

There is no doubt that such attempts to derive moral norms from nature are, as John Stuart Mill long ago pointed out, extremely problematical (Mill, 1885). This kind of ecologism is a valid and important target. But this naturalistic ecologism isn't generally characteristic of radical ecology. Indeed, it has more in common with the scientific/managerial approaches with which Haywood contrasts ecologism; those who, like Garrett Hardin, the Club of Rome Report, and so on, have advocated institutional attempts to impose population control. It is certainly true that some environmentalists associated with the Sierra Club (apparently including Dave Foreman) have advocated immigration controls into the United States on ecological grounds. But other environmental activists have been vocal in their opposition to such "use of slippery slope terms like 'carrying capacity' . . . population control always comes down to two questions: which populations and control by whom?"[14] They point out that "[e]nvironmentalists are fond of saying that pollution knows no boundaries, so it is ironic that some . . . could argue for greater control over these *artificial political boundaries*." In general, radical environmentalists are actively suspicious of anything that smacks of a naive naturalism. It is neither accurate nor fair to ignore antinomian strands of environmental activism and even Haywood has to admit that deep ecologist Arne Naess, the founding influence on deep ecology, dismisses such scientistic naturalism as an "excessive universalisation or generalisation of ecological concepts and theories" (Naess in Haywood, 1994: 19).

It is ironic that Haywood's attempt to promote a "progressive" humanism should leave him in a position close to right-wing critics of deep ecology like Luc Ferry. Both regard environmental critiques of modernity as indicative of a return to the rigors of natural necessity; as advocating society's submission to the laws of nature. Both authors also see this as threatening our current post-Enlightenment ideas of social justice. Here though, the resemblance ends, since Haywood appropriates certain aspects of radical environmentalism for his own purposes, translating everything in radical ecology that meets his approval into the language of his own particular brand of ecological humanism.[15] The rest, including the "romantic," "utopian," and "religious/spiritual" aspects of deep ecology, are dismissed as wrongheaded or superfluous. By contrast, Ferry regards these supposedly irrational elements as central to radical environmentalism's worldviews but, because of this, finds them all the more disturbing.

While Ferry at least recognizes that "American, German, and French versions of deep ecology are[,] in all cases . . . a matter of questioning the modern tradition of legal humanism" (Ferry, 1992: 73), he believes that this heralds a new "cult of life," an antihumanist and therefore "inhuman" religion of nature worship. The rule of humanist law is replaced with a "great messianic master-plan" (Ferry, 1992: 139) and a militant "religious fundamentalism" (Ferry, 1992: 134). This irrational attack on the institutions of modernity can only lead to either extreme left-wing or right-wing versions of totalitarianism; deep ecology "continually hesitates between conservative romantic themes and 'progressive' anticapitalist ones." Its schizophrenic attack on modernity means "some of deep ecology's roots lie in Nazism, while its branches extend far into the distant reaches of the cultural left" (Ferry, 1992: 90).

Ferry paints a picture of an ecological Nazism, a "New Ecological Order" that will rely on coercion to institute its environmentalist agenda. It will not be democratic; the choice is a simple one "[b]etween barbarism and humanism" (Ferry, 1992: 151). He engages in an often surreal polemical defense of rationality and humanism that would be funny if it did not signify a serious and completely unjustifiable attempt to paint any concern for nonhuman entities as an expression of fascistic religiosity. For example, the battle between "zoophiles" and "zoophobes" is apparently exemplified in the fact that something like half of one percent of all medical emergencies are caused by dog-bites! He thinks that the fact that people whose dogs savage small children usually get only a small fine is indicative of placing our love for animals over humans—but entirely forgets that the dog concerned was trained by and for these humans and is itself often put to death. He simply assumes despite the considerable literature in this area that animal rights and environmental issues are identical, exhibiting an almost total ignorance of the nature of the debates he enters. More importantly he completely fails to take environmentalism's antinomianism into account.

His arguments that Nazism and ecology share much in common is the intellectual equivalent of claiming that because Hitler was a vegetarian all vegetarians are Nazis. It is entirely facile to state that "an interest in nature, while it may not imply a hatred of men *ipso facto*, does not exclude one either" (Ferry, 1992: 92–93). No one should underestimate the possibility of right-wing groups utilizing environmental rhetoric to their own ends, but environmental discourses are no more open to this fascistic misappropriation than those of humanism and rationalism. Indeed, a good case can be made that Nazism entailed the extension of those (modern) principles, such as instrumental rationality, more than it did their rejection (Marcuse, 1998; Bauman, 1989). If anything, radical ecology's profound distrust of authority of any kind and its insistent emphasis on tolerance, the decentralization of power and learning from other cultures make it much less likely to be subject to this kind of misappropriation.

While Ferry and Haywood both illustrate the potential dangers in an institutional and authoritarian form of environmentalism, they both fail to take the anti-authoritarian character of radical strands of environmentalism into account. Radical environmentalists do not seek to impose a monolithic "ecological order" but to develop lived ethics that express something of the innumerable potential variety to be found in our relations to our environs. There is a world of difference between reading nature and society in terms of the nomothetic model that characterizes modern bureaucratic regimes and using our lived experiences of nature's profuse variety to illustrate just how bereft of beauty and feeling our legal/bureaucratic social world is.[16] Radical ecology does not want to make humanity subject to natural laws but to allow us to become subjects within a "natural" world.[17]

This nomothetic approach entirely fails to capture the individual phenomenology of the environmentalist's relationship to nature, replacing the heartfelt immediacy of ethical concern with an abstract post hoc rationalization; an unqualified "love of the land" with a set of codified legalistic principles.[18] The give and take that characterizes radical ecology is reduced to a series of prescriptions, an acontextual rubric designed to be worked by those who keep their distance from particular places and circumstances (Smith, 1997). The individual no longer has a moral "free range" but is caged within a legal framework based on somebody else's "reason." This rubric's supposed universal applicability contrasts completely with radical environmentalists' emphasis on the importance of gaining a sense of particular places (Snyder, 1984). This, after all, is why the particular trees, like Newbury's granny ash or

Merrick's favourite Mary Hare north tree have particular ethical significance for the protesters.

This returns us full circle to the question of antinomianism, which, if my characterization is at all accurate, rejects political-administrative authority, refuses to accept the external imposition of moral norms, questions reason's impartiality and instead emphasizes the inspiration of the free individual, liberty and love. Far from seeking legal reform it is more interested in turning the moral world upside down. In this sense we need to take seriously the claims of Robyn Eckersley and Andrew Dobson, who, despite their own differences, all argue that the "closest approximation . . . to the centre of gravity of a Green sustainable society . . . [is] the so-called anarchist solution" (Dobson, 1990: 83). Yet this is precisely the solution that many writing on environmental issues from within institutional parameters choose to ignore.

The Natural Contract

I want to conclude this chapter by making a few observations on one particular recent attempt to extend a nomothetic model to our natural relations—Michel Serres' *The Natural Contract*. Serres notes the close relationship between law and knowledge, between "justice and justness (or accuracy), reason that judges and reason that proves" (Serres, 1998: 65). These two rationales coevolve, mirroring each other's forms, but also competing against each other. "Does knowledge really parasitize the law, with which it chooses to maintain a relationship? But of course: it imitates it, mimics it, theorizes its form, refines it, and finally fights it to the death" (Serres, 1998: 65). The eventual outcome of this stormy relationship in modernity was the increasing dominance of science. "Science takes the place of law and establishes its tribunals, whose judgements, henceforth, will make those of other authorities seem arbitrary" (Serres, 1998: 86). This situation has lasted for three hundred years since Galileo showed, despite the efforts of the papacy and the old Roman law, that the Earth moves. Now once again, Serres argues, the Earth, that *de jure* and *de facto* ground of our being moves—only this time the changes are the consequences of the predominance of scientific rationality, for example, global warming. The solution, Serres argues, is to bring law and science back into a dialogue with each other over nature, since both share a concern with the need to maintain a balance. The scales of justice might complement those "laws of nature" that, we are told, "almost always come down to expressions of equilibria. . . . We could call them, literally, laws of justice" (Serres, 1998: 87). Equity, Serres argues, demands that we give something back to nature for that which it gives us, the "principle of reason thus consists in the establishment of a fair contract . . . with nature" (Serres, 1998: 90).

Serres' argument is obviously intended to be poetic as much as philosophical. The connections he makes between law and science are more tenuous than those that might be established by examining, as Weber did, the reflective operation of their particular rationales, the very real ways in which their forms mirror each other in modernity. Yet this poesy does not excuse the fact that his appropriation of a legal metaphorics is almost entirely uncritical. He simply assumes that the law is the natural home of ethics. He seems to forget that the principles of "[m]astery and possession," which he recognizes as underlying our "scientific and technological age" (Serres, 1998: 32), are also the same principles embodied in modern formulations of the law. He fails to question how and in whose favor the law's concern to "balance" powers actually cashes out, how its "justice" is administered, and how it constrains and disciplines those that are subject to it. He fails to recognize the many different models there might be of equity and equlibria—some owing little or nothing to the law's weight and measures.

The legalisitic model Serres utilizes is intimately associated with modernity. A reaffirmation of law's marriage vows with science would offer no respite to nature. The rule of law embodies and engenders a hierarchical relation between that social totality whose interests it supposedly serves and those who are subject to it. There is no reason to think that its relation to nature would be any different. The idea of a social contract, a mythic agreement to recognize the rule of law, is no less hierarchical because all citizens find themselves in the same boat as having, unbeknown to themselves, assented to their subjection. Those discourses that emphasize contractual agreements dispense with dependence upon the whims of a God who elects some to positions of absolute authority—the divine right of kings. But the mantle and authority of God is taken on by reason, which claims to express those principles to which all have supposedly agreed. This new nomothetic mythology may take the name of the people themselves as a justification for the statutory and juridical powers acquired and implemented by those who police society, but though the arbitrariness of such power is masked, it is no less real.

This contract, which is no less binding for its never having existed, is a legalistic rationalization intended to solve the problem of maintaining order in modern society. It is a myth fostering the efficient management of society, in Rousseau's own words, "to bring together what right permits with what interest prescribes so that justice and utility are in no way divided" (Rousseau, 1987, 49). The maintenance of order among the diverse interests and values found in modern society requires that the law be seen to take the role of an impartial arbiter, that it apparently divests itself of ties to particular sections of society (whether or not it actually does so).[19] The law's role thus becomes primarily functional; it too is dominated by instrumental concerns, paramount among which is to ensure the efficient running and continuation of the *current* social order. The law thus punishes those who disturb this order on the grounds that they harm "society." In Foucault's (1991: 92) words, "the injury that a crime inflicts upon the social body is the disorder that it introduces into it: the scandal that it gives rise to, the example that it gives, the incitement to repeat it if it is not properly punished, the possibility of becoming widespread that it bears within it."

This of course bears directly on radical environmentalism, since its antinomian aim is precisely to subvert the current damaging social order, to expose the problems that arise from making efficiency and the maintenance of social order "laws unto themselves." There is thus a real irony in Serres' attempt to extend the idea of a social contract to that of a natural contract. As John Clark points out, although Serres' motives may be impeccable the "term *contract* bears . . . [the] burden of history," of its use as a "myth to justify political domination and economic oppression" (Clark, 1996: 118).

Although Serres' work points toward a re-evaluation of the relationship between society and nature, a recognition of the need for give and take, we need a genuine dialectic with nature rather than a *formal* "balance" or contract. As the following chapters argue, any balance must be achieved in a manner that avoids, so far as is possible, the constraints imposed by the scales of modern nomothetic justice.

6

Against the Enclosure of the Ethical Commons: Radical Environmentalism as an "Ethics of Place"

At the beginning of this book I argued that we must recognize the innumerable but nonetheless important ways in which the moral values of any social group are linked with that culture's prevailing "forms of life." Particular values only make sense when seen in the context of the patterns of interconnecting and overlapping social (and environmental) relations and practices that are the medium within which they are produced.[1] Put simply, how we live obviously affects what we value and how we come to value it and values also work reciprocally to inform how we live. This dialectical relationship, at what philosophers refer to as the "normative" level is, I suggested, also repeated at the meta-ethical or theoretical level.[2] Theories of value are never, whatever they may claim, entirely abstract or context free. The *form* as well as the *content* of those ethical discourses that attempt to define and shape the moral field are also inevitably products of particular times and places.

I want now to begin to draw together some of the strands of the previous chapters, that is, of rationalization, the exclusion of nature, of the anthropic bias in production, environmental antinomianism, and so on, by illustrating the differences between the *moral architecture* of modernity and the anarchic, postmodern *ethics of place* that characterizes radical environmentalism.[3] The differences between these distinct moral fields will be expressed in a spatiotemporal metaphorics, a discourse of relativity, proximity, dimensionality, distances, volumes, velocity, and so on (Keith & Pile, 1993; Lefebvre, 1991; Silber, 1995; Soja, 1990). However, when speaking of a spatial metaphorics, it is important to dispel any notion that references to "moral spaces" are meant *only* metaphorically—they are not. Moral fields are a socially produced reality that, in Edward Soja's terms, "both reflect and configure being in the world" (Soja, 1990: 25).[4] Our ethical architecture forbids or facilitates behavior just as effectively as walls, windows, or doors. This work is metaphorical only in the Nietzschean sense that all language is to some degree metaphorical, that the truth is "a mobile army of metaphors" (Nietzsche, 1966: 46).

An "ethics of place" (i.e., of environments) reconnects moral and physical spaces in such a way as to subvert our present ethical agendas. This project has obvious affinities with the rejection of dominant ethical frameworks by feminists like Carol Gilligan who sees "in women's thinking the lines of a different conception [of ethics] grounded in different images of relationships and implying a different interpretative framework" (Gilligan, 1994: 208). Gilligan reacts against the developmental models of psychologist Lawrence Kohlberg, who envisages an ontogenic series of stages passed through by each child on its way to ethical maturity; a maturity that he equates with attaining the ability to apply abstract rules and principles to moral problems. Gilligan rejects Kohlberg's systematization of ethics, his production of a codified hierarchy of progressive stages each successively more universal and abstract, each less contaminated by messy contextual realities. She recognizes this for what it is, not an objective scale, but the sanitized product of an institutional context. Kohlberg's conceptualization of ethical development is symptomatic of its sphere of operation. It is a sterile masculinized parody of moral experience, a tool for subjecting morality to clinical trial and expert diagnosis, a method for measuring ethical conformity by positing as universal and natural an atypical, gendered, and historically particular set of criteria.[5] Gilligan contrasts Kohlberg's work with an "ethics of care" that does not rely on any rational or procedural universalism but calls attention to heartfelt feelings and their contexts. Her project too can be seen in a spatiotemporal light as (re)asserting the relevance of *situation* and *proximity* in ethics. She "calls into question the values placed on *detachment* and *separation* in developmental theories and measures" (Gilligan, 1994: 214). Her projected ethic does not simply seek to redress a perceived imbalance between reason/emotion or abstract/concrete by adding weight to neglected elements of ethical experience. Rather, she wishes to subvert the sphere of ethics as it is now and reconstitute it with different interests at its heart. Her project is revolutionary rather than reformist and both opens and articulates a different moral space; it sustains and is sustained by a different moral field. An environmental ethics of place must similarly revolutionize and reconstitute modernity's boundaries and problematics.

While the following chapters will attempt to spell out the nature and implications of this moral problematic in more detail, my present aim is merely to sketch the broad outlines of the ethical conflict between the principles underlying modernism and environmentalism. This conflict is inevitable since modernity constricts and diminishes the space for nature's own productive activities. In Lefebvre's words, nature is disappearing, "becoming lost to *thought*. . . . Even the powerful myth of nature is being transformed into a mere fiction, a negative utopia: nature is now seen as merely the raw material out of which the productive forces of a variety of social systems have forged their particular spaces. True nature is resistant, and infinite in its depth, but it has been defeated, and now waits only for its ultimate voidance and destruction" (Lefebvre, 1994: 31). Yet this defeat is not, as Lefebvre suggests, absolute. There is still a hope that nature might yet resist modernity's transformational intentions and come to be recognized as more than mere resource. First though we must grasp something of the spatiotemporal structuration of modernity.

Modernity: A Diagnosis

A catalogue of recent structural changes associated with "modernity" might include the industrialization of production, demographic shifts to urban centers, the development of global systems of economics and mass communications, the rise of nation states, the growth of bureaucracy, and so on. As novel institutional structures have arisen—banks, the stock exchange, transnational corporations, and so on—others have fallen from grace. Each change brings in its train new requirements—for a standing army and police force, centralized government, systematic education, and so on. Each change also heralds new ways of thinking, behaving, and valuing. The increasing rate at which modern society grows and changes, and the increasing social fragmentation that ensues means that fixed values and ideals are of only fleeting relevance and seldom likely to be shared with one's contemporaries. Attempts to impose given ends, however egalitarian they might be, are always compromised by the unceasing and accelerating motor of change. Chaos and confusion mark "modernity" as a context wherein the social being (verb) of the earth's inhabitants is repeatedly disembedded, where previously held certainties are overthrown and the ground is constantly swept from beneath our feet. Marx and Engels elucidate modernity's promise and curse thus: "Constant revolutionising of production, uninterrupted disturbance of all social conditions, everlasting uncertainty and agitation distinguish the bourgeois epoch from all earlier ones. All fixed, fast-frozen relations, with their train of ancient and venerable prejudices and opinions are swept away, all new formed ones become antiquated before they can ossify. All that is solid melts into air, all that is holy is profaned" (Marx & Engels, 1977: 36–37; Berman, 1991). The very magnitude and rate of these changes forces upon modernity the recognition of its own uniqueness and its acceleration produces new patterns of shared experience even if these can only be described in terms of alienation or anomie.

One can imagine the coming-to-presence of modernity as a process of metamorphosis, as a transformative restructuring and reorganizing of the world according to new principles, interests and needs. Caught up in its own ideological presuppositions, such change portrays and justifies itself as "progress," as a directed linear process that is both inevitable (natural) and a matter of constant improvement over a receding past. This ideology of modernism and progress acts as a chitinous carapace—a crystalline shell that shields the process of change from critique. Modernity is, thus imagined, a *flight* from the past toward an uncertain future, a future portrayed by its proponents as the airy transcendence of our earthbound and grublike past existence. But the promised future never arrives, each ecdytic moulting reveals not an imago with iridescent wings but tarnished images of more objects that might yet be produced, processed, and consumed. Change itself becomes modernity's principle of organization, continual growth and movement its only recurring theme, increasing speed and efficiency its only measures of success. Progress refers only to a movement where what comes later is regarded by definition as better than that which went before. While modernity comes to objectify in its very structures a conception of a *world without ends*, of pure blind instrumentality, environmental critics point out that such a conception is more likely to entail *the end of the world*.

Ironically, then, "modernity" is most easily recognized not by what it is at this moment, for this never remains constant, but by its wholesale orientation toward what it might be. According to Goran Therborn, modernity "is characterized by a discovery of the future more than by a valorization of the present; that is, the discovery of the future as an open, *unbuilt* site never visited before, but as a place reachable and *constructable*. The present is the beginning of the future, rather than an extension, a propagation, a repetition or a decay of the past" (Therborn in Featherstone, Lash & Robertson, 1995: 71, my emphasis). But we must go further, for this obsession with futurity is not envisaged teleologically (as all previous futures were), it is a future without any specific end in sight. We inhabit an epoch where the process of getting to a future, *any* future, has become an end in itself; we run forward but cannot, or do not, look where we are headed or on what we tread.

In short, though lacking a unified vision of its own present and its future possibilities, modernity comes to rely upon and be dominated by the principles that seem to best express its own situation, those that also claim to eschew the promotion of particular ends. Thus its one-dimensional obsession with ever-increasing rates of change and growth generates tensions that constantly fracture existing social patterns and spawns a society of alienated atomistic individuals subject to an *instrumental rationality*. This in turn generates the conditions addressed and expressed by modern forms of ethics.

The Enclosure of the Ethical Commons

Between the eighteenth and nineteenth centuries the common lands, permanent pastures and "wastes" of England, traditionally home to a largely uncodified system whereby people enjoyed access to seasonal grazing, fishing, peat cutting, firewood, and so on, became subject to Acts of Enclosure passed by Parliament and the courts.[6] They were transferred from open spaces into discrete areas bounded by fence or hedgerows, from places where social and economic practices were determined by tradition and local agreement to private domains for the privileged to own and exploit. Whole village communities were transformed by these acts carried out in the name of economic and bureaucratic efficiency. For all the explicit political hierarchy of feudal society, the common lands were usually locally run with minimal outside interference, and for this reason they varied considerably, reflecting local environmental and sociopolitical histories.[7] The real "tragedy of the commons" was that a relatively adaptable and stable system of subsistence and a set of values tailored to particular places became subject to commercial and managerial control at a distance. This is not to idealize the obvious injustices in the feudal system but to emphasize modernity's tendency to impose change, centralize power, and marginalize alternatives.

This process of enclosure was not carried out without opposition, and the new ethical ideologies that accompanied it were recognized by many as inadequate substitutes for radical political change. Thus the Diggers of the English Revolution like Gerald Winstanley were not to be satisfied simply by the apportioning of extra "rights" but sought instead a fundamental change of the system of land ownership.

> Common ownership of the land was for Winstanley the basis of individual freedom and social equality. Freedom, he insisted, could not be defined in terms of specific rights or privileges, as exemption from certain restrictions or in its application to particular groups and classes; for that would be to place too narrow and limited a definition of the concept. For Winstanley, its essence lay, above all else, in the establishment of those social conditions that would permit the free and adequate development of every individual.

> Freedom became a function not of particular political institutions or rights, but of the fundamental nature of the social order." (Petagorsky, 1940: 183)

An ethics of place might make explicit the links Winstanley recognized between social and economic relations, land, locality, and ethics. The parceling up and privatization of land goes hand in glove with the reconstitution of the ethical sphere along similar lines. For this reason the term "enclosure" might serve to express the incarceration of moral spaces within conceptual and practical frameworks that codify and embody ethical values in modern philosophical discourses and social institutions.[8]

Today ethics is dominated by two abstract theoretical paradigms: utilitarian and rights based. Though often regarded as opposing tendencies, they nevertheless resemble each other insofar as they both embody and reinforce the principles of that modern society with which they coevolved. Both ethical paradigms entail the systematization and institutionalization of procedures of *deferral* and *distantiation*, that is, deference to authoritative bodies whose objectivity is supposedly guaranteed and validated by their very distance from, and nonparticipation in, the activity concerned.[9] Both use explicit and abstract formulae and arguments that are supposed to float free of particular contexts. In the case of utilitarianism, administrative decisions are given that illusive purity associated with abstract mathematical manipulations. Rights based systems bring morality before a higher court in whose eyes all are purported to be equal; a body of "impartial" arbiters who confirm or reject behavior in terms of its conformity to abstract principles, rules, and regulations. Each encloses what it regards as morality within a codified and explicit rubric, whether this takes the form of a consequentialist calculus of benefits and losses, or of written law or predefined procedures. Both portray themselves as attempts to make morality *accountable* to all, and these accounts must therefore be in a currency acceptable to all, that is, they must be cashed out in terms that all would supposedly understand and agree with. Thus ethics too, it is claimed, must be ordered by that supposedly universal and value-free human feature "reason." In this way cold reason comes to pass judgment upon a moral field that might at first seem to be the natural habitat of its own self-decreed enemies—emotion and tradition. This call for accountability in the face of fragmented and disparate values is then both the source of, and the rationale behind, ethical discourse's increasing abstraction from specific environments and its increasing subservience to a reasoning that supposedly eschews all contextual bias.[10]

Modern morality becomes subject to those who claim to impartially represent people in general. In other words ethics comes to be administered by those professionally "disinterested" bureaucrats who constitute and uphold those institutions and regulations misleadingly referred to as "representative" government and the "common" law. Bureaucracies act, as Weber recognized, as "*the* means of carrying 'community action' over into rationally ordered 'societal action' . . . as an instrument for societalising relations of power" (Weber, 1964: 228). Thus morality comes to find itself deferring to the upper echelons of a hierarchy that, despite all the accumulated evidence to the contrary, continues to portray itself as encapsulating the interests of all and to justify its *private gains* by a rhetoric of *public service*. The ethics of this system enclose the moral world according to principles of socialization rather than sociality, that is, they are concerned with interpellating the individual into an already given social framework and promulgating a top-down set of values to be internalized rather than letting values form and operate from the bottom up by communal participation.[11]

In its raw Benthamic form utilitarianism is just about as crude an expression of the principles of an emergent modernity as one could imagine. It eschews specific social ends and refuses to envisage or promote any particular social order or intrinsic values. Utility is a reified measure of ethical "progress," the calculation of which becomes an (anti-)social end in itself.[12] It is the objectification of an (almost) entirely instrumental ethical space. Accountability is its very raison d'être. Here all values are reduced to a single quantifiable currency, that of "pleasure," and ethics becomes a mere felicific calculus. Utilitarianism represents, in more than one way, the economisation of the ethical sphere—its subsumption under and realignment with the principles of quantification, private ownership, and exchange. Utilitarianism and neoclassical economics alike are concerned with the satisfaction of individual desires and the measurement of individual preferences. Both regard society as nothing more than the sum of the individuals who compose it.

In utilitarianism, ethical progress is ensured not by the *trading of* economic goods between individuals but by a *trade-off* of moral goods administered by moral accountants. Bentham differs from free-market proselytes only in the extent to which, as a lawyer, he agrees with rights theorists in promoting the role of a professional legislative bureaucracy to order society, in his case, in conformity with the principle of utility. Where Adam Smith thought all could be left to the "invisible hand" of the market, Bentham defines a space watched over by the invisible eyes of the public administrator. Bentham's perfect society cannot be achieved simply by each following their own self-interest, but requires that each be made aware of and internalize the gaze of the generalized other. Bentham's ethical space is objectified and given concrete form in the architecture of his Panopticon, a multipurpose institutional building designed to facilitate the efficient control of its occupants, whether they be prisoners, hospital patients, or schoolchildren (Bentham, 1995).

For many commentators "rights" serve to guarantee the individual's civil liberties against both the whims of despots and the relativizing and reductionist tendencies of utilitarianism, but deontology too expresses its modernist heritage. Both utilitarian and rights-based systems alike bear witness to the privatization of the ethical, the parceling up of a communal field into individually owned packages of duties and obligations. Indeed, to some extent, each can be seen as acting in compliance with the other, rights providing the institutional stability required to allow the consequentialist trade in values to flourish.

As Marx notes, even in its most radical forms deontology embodies the values of an emergent capitalist economy. Paradoxically, even in the midst of revolutionary France, the Declaration of the Rights of Man "soberly proclaims the justification of egoistic man separated from his fellow men and the community" (Marx in Waldron, 1987: 147). For Marx the discourse of rights presupposes as natural the existence of that alienated individualism that is actually an artifact of capitalism. Buchanan notes that Marx "condemns the very notion of a (legal) right as being an artifact of defective modes of production and believes that such a notion will become obsolete when a superior mode of production comes into being" (Buchanen, 1982: 68). Marx, like Winstanley before him, does not regard a generalized notion of equality as distributive justice as either necessary or sufficient to characterize a genuinely communist society. The real needs of people—who insofar as they are individuals are always different (i.e., not equal)—cannot be met by systems that treat people as standardized abstract elements and regard justice as an administrative problem rather than as a matter of radically transforming society. The famous call "(f)rom each according to his ability, to each according to his needs" (Marx in Fischer, 1970) is not a call for equality but for *community*, which is something entirely different. The focus on rights is an ideological consequence of the social sphere engendered by capitalism and for those seeking radical change it is, in Marx's own blunt phraseology, "obsolete verbal rubbish" (Marx in Waldron, 1987: 135).[13] Similarly, as the last chapter showed, Foucault argues that the legal institutions like the courts that embody the deontological ethos are far from being "the natural expression of popular justice" (Foucault, 1980).

In Marx's terms the discourse of rights serves to enclose moral debate within the ideological sphere of capitalist society, it facilitates a limited conversation about proprietorship, propriety, and the distribution of goods while discounting alternative discourses that situate themselves in opposition to this very social order. The prioritizing of rights is nonsensical out-with a system that recognizes private property, celebrates an abstract individualism, and centralizes decision making within a legislative and administrative bureaucracy. Yet via cultural reinforcement and repetition these boundaries on moral thinking come to be seen by theorists and practitioners alike as natural and immutable. They hedge about the moral field in such a way that even practical action is orientated toward, enshrines, and thus reproduces the limits and possibilities afforded by these abstract codified principles and rules (Smith, 1995). Utilitarianism and deontology alike operate as an ideological horizon that those committed (whether consciously or not) to modernity's aspirations regard as the natural limit to the moral sphere. All attempts to confront this ethical order are, for them, unrealistic if not strictly unthinkable.

The end result of this process of rationalization is the complete alienation of the person from their moral potential, from the values that would make them real rather than abstract and impoverished individuals. Those moral values that do survive in the midst of this abstract codification and pragmatic instrumentalism find that they are undermined and reduced to echoes of their former selves. They become moral simulacra divorced from the context that gave them birth and relevance and inserted into systems that are now mere formal procedures designed to keep society running smoothly on its unquestioned course. The moral sphere has become subject to the pragmatic needs of government and the global capitalist economy. We have a tame ethics, an ethics of control that *encloses* the moral field via a process exactly analogous and closely allied to the enclosure of Britain's social and agricultural spaces.

A common thread unites those who, from Winstanley through Marx to today's environmental activists, oppose the enclosure of the moral field within a discourse that refuses to question its own social origins and legitimacy. As the last chapter showed, the recent occupation of St. George's Hill by activists, the site of the original Diggers settlement, made an explicit connection between historical and contemporary protests against the enclosure of the land and the moral enclosures of the Criminal Justice Act 1994. In both cases the real issue is not one of conflicting rights of access to and ownership of private property but of disagreements about the economic and political system that gives rise to this kind of proprietorship itself. While radical ecologists do not necessarily all call for the abolition of private property, they do see the land as an inalienable home for those who inhabit it and not as an object to be bought and sold for economic gain. If this kind of protest is to express the issues from the protester's perspective, it requires an ethical vocabulary that does not find itself tied to a discourse that at least tacitly operates in support of the current social and moral order.

The Ethical Architecture of the "Open Road"

How then might a spatiotemporal metaphorics help us to gain a different understanding of the ethos of radical environmental action? We might better understand what has *gone wrong* with modernity by envisaging an ethical language capable of expressing something of the moral disquiet present in environmental protests. An "ethics of place" might give voice to concerns that are otherwise marginalized or misrepresented in hegemonic discourses.

Roads provide a perfect example of that "industrialization of time and space" that has characterized modernity (Smith, 1998). The modern road is a "devitalized" and abstract space torn from its surrounding social and environmental matrix. It compresses the natural and social complexity of the journey into a vacuous experience of instrumentality, a one-dimensional straight line. Wolfgang Schivelbusch's description of the railway journey on the nineteenth-century psyche is equally applicable to modern roads. "What was experienced as being annihilated was the traditional space-time continuum which characterised the old transport technology. Organically embedded in nature as it was, that technology in its mimetic relationship to space traversed, permitted the traveler to see that space as a living entity" (Schivelbusch, 1986: 36).

In Marc Augé's terms the modern road might be seen as a "non-place"—somewhere that serves merely as a conduit with no existence in its own right (Augé, 1995). No matter what the width of its carriageways, no matter how many lanes it has, nor how much countryside it consumes, it remains fundamentally one-dimensional, its traffic streaming along a single axis. What is more, every individual's relation to this "place" is uniform—functional and unencumbered. In this, of course, it also epitomizes the mass-produced and compartmentalized "individualism" so characteristic of modernity. This was not always so. In the not so distant past, the road was a place of encounter (can one even imagine today a modern Canterbury Tales?), a space to be inhabited, linked as it was to local traditions, practices, and conditions. Each bend had a history, every twist its own special rationale. Even the most extreme celebrants of modernity have recognized these changes though they did not mourn their passing. Thus in 1924 Le Corbusier could remember the coming of

> the autumn season. In the early evening twilight on the Champs Elyses it was as though the world had suddenly gone mad. After the emptiness of the summer, the traffic was more furious than ever. Day by day the fury of the traffic grew. To leave your house meant that once you had crossed the threshold you were a possible sacrifice to death in the shape of innumerable motors. I think back twenty years, when I was a student; the road belonged to us then; we

> sang in it and argued in it, while the horse-bus swept calmly along [. . . and yet Corbusier] was overwhelmed, an enthusiastic rapture filled me . . . the rapture of power. The simple and ingenuous pleasure of being in the center of so much power, so much speed. (Le Corbusier, 1971: 3)

The motorist pays for smooth speed enhancing surfaces by having to focus on the road ahead; the faster they go the less they see of the places en route. On the road all life becomes functional in the narrowest sense. The road is a space dominated by considerations of utility; built and managed in order to increase traffic flows, its primary function is to transport people from one place to another as efficiently and quickly as possible. Twyford Downs in Southern England, its ancient tracks and archaeological sites, its rare habitats and its tranquillity, was destroyed to cut eight minutes off the journey time from London to Southampton. The Newbury bypass will save just two minutes on the average journey.

The road, like all instrumental structures, is supposedly a neutral facilitator of individual aspirations. It has no in built *telos,* but merely serves to increase trade and travel, helping all and sundry on their various and unspecified ways. But instrumental rationality always conceals a partisan politics behind its neutral mask. Instrumentality inevitably imposes its own constraints upon those living in its shadow. Once regarded simply as a *means* of transport, roads and their allied industries have become a central plank of what Mrs. Thatcher famously described as "Britain's great car economy." We find that we are forced to restructure our lives to fit the needs of that which was supposed to serve our interests. One might say, without exaggeration, that the modern road objectifies and embodies that process of rationalization so forcefully described by Weber, though perhaps even he would have been surprised that his "iron cage" would take the form of an upholstered steel box on wheels.

Of course, there is a mythology of the open road, empty and stretching out to the distant horizon, that fits perfectly with modernity's aspirations. In the genre of the road movie the road is a symbol of unconstrained freedom.[14] But this myth exists now only in the dusty memories of the past, in Hollywood or more commonly in the flash dishonesty of the car advertisement. It says nothing of a landscape dissected by tarmac, of asthmatic children, the flattened and bloody corpses of badgers and birds strewn in rubbish-filled gutters, of smog, gridlock and road rage.

And what of deontology? Is the road purely utilitarian space? Here, as elsewhere, we can see that the incorporation of deontological concerns facilitates rather than opposes considerations of utility. Again "rights" and legal codification in general serves to provide a framework within which utility can work more efficiently. The most obvious examples are things like the Highway and Greencross Codes, a series of written rules that serves to reproduce, via the inculcation of a modern *habitus*, a set of relations and values to which all road users should defer.[15] Primary among these is the need to recognize that the road is a space for vehicular traffic, not people. From an early age all children are inculcated with a necessary respect, pedestrians should always *give way* to automobiles. Such rules are mostly concerned with letting road users "go about their (and capitalism's) business," and are intended to coerce those who might obstruct their "rights of way" or infringe on their liberty to transport dangerous or unnecessary cargoes. For these reasons if no other the advent of recent road protests like "Reclaim the Streets" marks a radical challenge to the instrumental, one-dimensional, and codified ethos of the modern road.

Toward an Ecological Ethics of Place

In 1994 in East London the £250 million M11 link road ran up against the newly instituted state of Wanstonia, comprising those streets threatened with demolition by the encroaching asphalt. In Claremont Road the protesters and sympathetic residents occupied deserted houses and devoted their considerable imaginative energies to delaying the road in every way. In contrast to the linear one-dimensionality of the incoming road, this community *revitalized* the locale, planting trees, producing sculptures, turning old cars into home-made road blocks, and building tree houses. Both sides of Claremont road were joined by netting and rope-walks, and the houses connected by a network of tunnels at all levels creating a multidimensional space dedicated to slowing the pace of the road's construction. Hundreds of police and bailiffs struggled to clear the street inch by inch over a period of weeks as the protesters nonviolently resisted their advance, some even partially burying themselves in concrete blocks. In George Monbiot's words "the creative impulse which erupted in Claremont Road was one in which everyone participated, rebuilding the world as it suited them. Dispossessed city youths knocked holes in walls between the houses and established tracks across the rooftops, building a communal space from the architecture of alienation" (Monbiot, 1994).

Similarly in Pollok Park, Glasgow, the scene of the M77 extension, protesters created a semipermanent encampment incorporating platforms and tree houses, an alternative university, and a plethora of sculptures invoking appropriate Celtic symbolism. Environmentalists and local residents both combined to defend the newly formed "Pollock Free State" against the contractors Wimpey, who showed their appalling ignorance of the protest's ethical dimensions by trying (unsuccessfully) to buy the sculptures to enhance the new road.[16] This pattern of protest is being repeated in ever more imaginative ways all over Britain and in many other parts of the world.[17]

The clash of ideologies and forms of life present in these protests could hardly be clearer.[18] Nor should it surprise us that this clash is present in every aspect of the protesters' camps, from the multidimensionality of their tree houses to their neotribal communalism, their dress, their language, names, and values. Contrast the uniformed and formally equivalent security guards and police with the anarchic celebration of difference by the protesters, made graphically clear in photographs of the arrest of protesters in the guise of "Daisy," the pantomime cow at Newbury. Contrast the police, identified by number, with the contextuality of the protesters' names, taken from resident species like Badger, or the Donga tribe (named after the ancient trackways partially destroyed by the road at Twyford Down in Southern England).

We are witnessing a confrontation between two quite distinct ethical architectures, two different ways of shaping the ethical field. Where the roads' advocates couch their arguments and tactics in abstract terms of economic utilitarianism and legal rights, the protesters emphasize the specific ecology and ethos of the locality they defend, its uniqueness, special qualities, and associated traditions. To counter the frantic timetable and singularly destructive aim of the contractors, the protesters seek to slow things to a walking pace, a pace at which the multidimensional complexity of the locale they protect, whether urban or rural, can work its spell.

Lacking other effective linguistic resources the protesters will on occasion use an ethical shorthand that might, for example, speak of a wood's "right" to exist. This language has a tactical role in communicating the ethical dimensions of the protest. But this is not a call for the further institutionalization of morality. Indeed most protesters explicitly recognize that it is modernity's institutional culture and its requirement that one must defer to others in moral matters that has allowed us to come to treat nature as a mere resource. To take this talk of rights legalistically would be a mistake. Nor does the astuteness of protesters in using the law to their own tactical advantage imply a regard for and compliance with the institutional legislative structures that try to hedge them in. Many environmentalists are, like Marx and Winstanley before them, seeking more radical change.

As the last chapter illustrated, the ethos of radical environmentalism is often misunderstood precisely because those who comment upon it make little effort to step outside and reflect upon the dominant ideological presuppositions of modern life. They try to enclose and constrain the ethical wilderness by superimposing upon it an unsuitable linear logic. The environmentalist's advocacy is misconstrued, squeezed into frameworks that can only analyze their meaning according to principles that the environmentalist's practices and forms of life implicitly, and often explicitly, reject. The arrogant dismissal awaiting many environmental discourses is a result of an imposition of a literal reading, a reading that, by definition, applies a rigid grammar incapable of recognizing the unfolding and restructuring of language's use of metaphor, mimesis, and metonymy.[19] This reading also disenchants language and forces it to defer to those normative meanings prescribed by the modernist taxonomy of the world. This disenchantment, which eschews metaphor and allusion (or classifies them as fictive), is itself a product of a one-dimensional instrumentality. Perhaps, instead of trying to nail down meanings, one should let them settle like falling leaves. One must, as Wittgenstein emphasized "look and see" how language is used in its own context (Wittgenstein, 1988: para. 66). Perhaps then we need not a literal but a *littoral* reading, a reading that finds new possibilities in the flotsam and jetsam of modernity's passing wave, at the margins of society or the end of the road.

The vagueness and indeterminacy of many attempts by environmentalists to give voice to their concerns are not a sign of irrationality as some have claimed, but of a struggle to speak using a language that makes certain things difficult to say (and apparently even more difficult to hear) to a culture that regards this language as a neutral and transparent medium. A sympathetic reading shows that these formal interpretations rarely connect with the radical intentions or actions of their originators, an error on a par with interpreting Marx as being simply concerned with distributive justice.

No codified rubric can accommodate, nor any axiology contain, the ethos of radical environmentalism. The problem of how to express and communicate this radically different ethos cannot be solved by enclosing it within the logic of the system it seeks to subvert. The ethical space opened by environmentalists has a new vitality and like the climbing tendrils of the bindweed and the anastomizing rhizomes of the honey-fungus its effects move between the flat surfaces presented by modern life, creating cracks and fissures through which we can glimpse the existence of an/other possible way of life.

Environmental Ethos and Ecological Community

The clash of cultures and of ethical values exposed in conflicts over road building could hardly be clearer. This, I have argued, is because radical environmentalism is engaged in a fundamental critique of modernism; its alternative culture challenges modern life to its very core. The ethical and social architectures expressed in and engendered by the physical architecture of cities and roads are poles apart from the anarchic creativity of the protesters' camps.

The road protesters' small-scale but multidimensional and symbiotic communities are not concerned with, and do not require, an ethics that functions as an administrative or regulatory tool. Rather, in rejecting modernity's destructive linearity, in breaking through the limitations and boundaries it seeks to place upon them, they release an anarchic creativity and reconstitute the ethical commons in a natural and highly politicized *situ*. The unfolding activities of these radical communities conjoin language, values, and practices, in a form of life that reconstitutes the relationship between nature and culture, an antimodernist form that rejects the monolithic logic of relations predicated upon control and consumption.

Such radical communality is not marked by formal codes and rules but by the attainment of a *modus vivendi*, a way of living that gives due, but not formally equal, respect and care to all its members. This is true whether such communities are envisaged in intraspecific terms, as Marx's communist utopia was, or ecological and interspecific terms. For radical environmentalism, the bounds of community do not stop at humanity, but include other aspects of our environs, whether animate or inanimate, with which we form relations of care and consideration. The only bounds on such communality are those recognized by the practical logic of day-to-day life and by the limits of our imaginations.

One example of an interspecific community is that "mixed community" of humans, bears, sheep, and wolves described by deep ecologist Arne Naess (1979). Naess uses the situation in southern Norway where bears and wolves live side by side with communities of humans and domesticated farm animals to illustrate the idea of a *modus vivendi*. There is inevitably some conflict of interests within this "community." Bears occasionally kill sheep, but farmers expect a certain level of predation since it is acknowledged that bears have to eat too. Only if the killings become persistent will any action be taken. However, there are no rules to dictate when and where a bear can be killed. All manner of factors might be taken into account. "What is his or her [the bear's] total record of misdeeds? How many sheep have been killed? Does he or she mainly kill to eat, or does he or she maim or hurt sheep without eating? Is particular cruelty shown? Is it a bear mother who will probably influence her cubs in a bad way? Did the sheep enter the heart of the bear area or did the bear stray far into established sheep territory?" (Naess, 1979: 237). One might think of other issues that could be pertinent, the severity of the weather and availability of other food sources, the degree of effort the farmer expended in protecting the sheep, and so on. In this way a complex discourse ensues that can take many varied factors into account, from the degree of recompense available, to the individual characteristics and habits of the bear concerned. No formal rules operate to dictate certain penalties; they are not needed. "The interaction between the members of the community is not systematically codified" (Naess, 1979: 238). and Naess thinks that given the problems that ensue on attempting such codification "it may be wiser . . . not to introduce the term 'right' in codification of norms concerning animal/human interaction" (Naess, 1979: 239).

This feature of Naess' approach is often overlooked by those who read into his references to biocentric equality an argument for extending legalistic frameworks to our natural relations, that is, as proposing the recognition and codification of a formal equality between species. But while Naess asserts in other contexts that one must "announce your value priorities forcefully" and start with "explicit rules" and "norms" (Naess, 1989: 68–69), these rules are not meant to regulate our relations to other species via a codified legal/bureaucratic system. They operate only as a loose framework, a statement that is intended to highlight the differences between deep ecology and those anthropocentric approaches that take no ethical account of other species at all. In other words, the argument that, all other things being equal, then no one species should be given preference over another, is not the kind of principle that can be made into law—for all other things are never equal. To set such an idea in stone, to formalize it, would inevitably lead, as Naess recognizes, to the many problems that analytic critics of deep ecology have noted, that is, that a formal biocentric egalitarianism would make human life unlivable, since every act from weeding the garden to vacuuming the floor would infringe on the "rights" of other species. But this formalization is not what Naess intends. Rather, he sees statements in favor of biocentric equality as ways of communicating a vastly different worldview to other people who, as yet, have not even considered it a possibility. Such statements are meant to act as *gestalts*—as reconfiguring people's way of thinking (Naess, 1989: 57–65).[20] It is meant to change our entire situation, to engender an entirely different way of relating to the natural world.

Making a statement in favor of biocentric equality is intended to act as a catalyst that might transform our patterns of thinking and our values. Suddenly, things are seen to "hang together" in a different way; we no longer place humanity *above* the natural world, but recognize that other species too are worthy of consideration. Such consideration does not require that we give them legal standing, it requires that we change the way we live. Our change in thinking will entail a change in our lives and be expressed in our care-full attitude toward other beings in our practical activities and discourses. This willingness to give consideration to others is exemplified in the interventions of radical environmentalists just as it is in the cultural traditions of the southern Norwegian farmers. In other words, for Naess, deep ecology is a philosophy that sets out to change the form of our lives. His principles are not meant to be blindly imposed on others but to open others' eyes to possibilities for developing their own novel ethical spaces, each suited to their own particular place in the world, each a form of "self-realization," but all united in rejecting the anthropocentrism prevalent in modernity. For this reason he refers to his own branch of ecological philosophy as "ecosophy T," noting that there might be many more varieties of environmental philosophy and forms of life—ecosophies A, B, C, and so on (Naess, 1989).

Naess' account fits the experiences of those who, through whatever experiences, have come to view the natural, environment differently from the norms dictated by our contemporary culture. The particular catalysts, the events, experiences, or arguments that can bring about such gestalt switches in our values and lifestyles might take innumerable forms—watching a lamb play while suddenly realizing one is eating a lamb chop—the all too familiar creaks and groans that presage the doom of a felled rainforest tree. The more intense the experiences, the more likely they are to signify a change. Many of those who begin by defending the environment for purely instrumental reasons have eventually found their whole worldview and values changed beyond recognition.

These changes often cannot be rationalized in terms that would make sense to those who play life by modernity's rules. To them the values that motivate environmental protesters are bizarre and perhaps even tainted with a dangerous irrationality. But, as Kate, one of the Newbury protesters says, "I am starting to forget myself why I first went down [to Newbury], but I know it was rationalized by well thought out arguments against the Car Culture. Now these arguments have been replaced by a belief system, an irrational commitment to the land, to the trees and to the people who fight to save them" (Kate in Merrick, 1997: 128).

Her fellow protester Helen describes the depth of these feelings and the changes in perspective that they entail. "I sat at the edge of Snelsmore and cried, I cried like I've never before cried in my whole life. . . . Slowly, totally, I released all the pain, emotions, terrible pain I had accumulated whilst watching the trees of Newbury being hacked to pieces and ripped out of the earth, watching them *kill* my home, *our* home, our earth as we know it. And still I cried with an energy and an agony that could not have come from me. It was like the earth was crying through me, I was releasing the bottled-up pain of every tree on the route, felled and standing. The earth was crying *through me*" (Helen in Merrick, 1997: 129).

Discourses of formal rationality and rights are incapable of expressing such depth of feeling for the environment or the values that arise from an ethically and emotionally charged dialectic with nature. The feeling of being close to nature simply makes no sense in those discourses that would pull apart, analyze, define and regulate both the moral field and the environment. For this reason, protesters find their concerns dismissed, marginalized, or even criminalized and must search to voice their concerns in other ways. Yet for all this they will not go away. The crass superficiality, the uniformity and one-dimensionality of the modern world cannot even begin to replace the diversity and beauty to be found and experienced in being party to the wider, natural community and of communing with that nature. This heartfelt ethics cannot be learnt by rote or dictated by rules but depends crucially on the development of a "feeling for life" and our place within it, that is, on the development of what might be termed an ecological *habitus* and an ethics of place. It is to this ecological habitus I now turn.

7

Thin Air and Silent Gravity: The Ecological Self and the Intangibility of the Ethical Subject

I hope that by now the pattern of this work has begun to emerge, that connections are forming which indicate the warp and weft of an alternative "ethics of place." This chapter will continue to weave together different filaments in an attempt to create a web whose fine tendrils might slowly gain a subtle and spidery strength through their reconfiguration of modernity's theoretical spaces.

Perhaps the most important strands of thought requiring further elaboration are those that link the "production of ethical spaces" with the "phenomenology of ethical feeling," that is, the twin aspects of morality's "objective" exteriority to individuals and its "subjective" interiority within individuals. How are the values that structure our relations to our natural and social environs reproduced and embodied in our selves, qua subjects, and our everyday practices? How do aspects of our environs, including human or nonhuman others, come to take a place in our hearts?

I intend to approach these issues through a critical examination of the writings of Louis Althusser, Judith Butler, Luce Irigaray, and Pierre Bourdieu, theorists who have, despite their different emphases and approaches, made the relationship between subjectivity and society central to their concerns. All consciously utilize a spatial metaphorics to express the manner in which the subject both incorporates and adapts prevailing social norms. This shared "spatiality" will enable the following chapter to link their analyses to Henri Lefebvre's (1994) concept of the "production

This chapter written with Joyce Davidson.

of space." The resultant problematic will then be used to provide the basis for reconceptualizing both objective and subjective aspects of ethical values in terms of the production of moral fields and an ecological habitus.

Space, Substantialism, and Relativity

Before proceeding to discuss Althusser and Bourdieu in detail I must return to the very idea of employing a spatial metaphorics, or perhaps more accurately a spatial *metonymy*, to replace more traditional theoretical approaches to ethics. As the previous chapter noted, it is all too easy to regard the use of spatiotemporal language in conceptualizing an ethics of place as merely allusive, that is, as figurative rather than expressive. Perhaps this is because we commonly tend to regard "space" as nothing more than that emptiness between substantial things, the vacuum between the stars, a distance to be traversed, a volume to be filled. We give our attention to those things that have a more tangible materiality, the hard surfaces that impede our "progress," the bricks and mortar of our habitation, the steel and stonework that give the impression of solidity despite their actual impermanence in the face of social change. This substantialism recognizes the import of only those things that manifest their reality in a manner that is immediately tangible to the senses.

This substantialist variety of materialism is closely allied with modernity's instrumental rationale as many theorists have noted. Max Weber, Martin Heidegger, and members of the Frankfurt School like Max Horkheimer and Herbert Marcuse each in their own way traced the rise of an instrumental/formal rationality that brings about the disenchantment [*entzauberung*] of the world, banishing all that is mysterious, mythic, or magical. The inauguration of this regime of instrumentality requires that all is made tangible to thought as well as deed. Clearly defined concepts cut the world "into pieces whose equality or difference we shall be able to evaluate, compare, reproduce" (Irigaray, 1985: 112); they are the tools that allow humanity to grasp, take hold of, and hammer the world into shape. For all these thinkers the modern world is redefined as a resource, a fixed and inert warehouse of materials awaiting manipulation, a standing reserve (*das Bestand*) that technology sets upon "demanding that it exhibit itself in its objectness and calculability" (Foltz, 1995: 12).

Having brought about the "disenchantment" of the world, instrumental reason concerns itself only with the concrete and practical. Since efficiency is as much a virtue in explanation as in other fields, it is characteristically loath to multiply entities beyond the bare minimum its Ockhamian ethos deems necessary.[1] Reason would prefer to explain the world in terms of things whose reality is readily apparent, things that exhibit a paradigmatic materiality, that are solid, fixed, bounded, isolable, and so on. Instrumental reason aligns itself with a Newtonian "solid mechanics," a "billiard ball" world where the material objects of analysis move in straight lines until they collide with others, a world whose laws of motion, of (equal and opposite) cause and effect, ensure the conservation of momentum. Once set in motion, this world functions smoothly with the efficient regularity of clockwork.

Tangibility becomes the touchstone of modern reality. Modernity's illusion is so palpable, so enticing, so complete, because reason conjoins instrumentality and substantiality. What reason cannot grasp we cannot use and what we have no use for simply has no place in our world—it is not and does not *matter*. There is thus, as Luce Irigaray has noted, a "complicity of long standing between rationality and a mechanics of solids" (Irigaray, 1985: 107), a complicity that denies reality to that which seems less tangible, which is less easily defined, more permeable, fluid, or reminiscent of those spirits that reason struggled so hard to exorcize. For Irigaray, this concentration on the rigidly defined and definite contours of solids is symptomatic of the hegemony of a phallocratic symbolic order that subsumes both women and nature within its own cultural logic.[2] This solid mechanics pervades every aspect of our lives, containing and constraining us within its instrumental parameters.

As modernity's offspring we even tend to understand our own identities and social relations in terms of substantialism's atomistic ideal/ideology of autonomous, bounded, individuality. The analysands of the modern social world are concrete and isolable individuals each on their own disparate trajectories, each with *particular* identities derived from their ownership of certain essential, quantifiable, and indefeasible properties. We are born under the sign(s) of one-dimensional *man*; *Homo economicus*, that self-contained and self-serving caricature of modern humanity, a parodic recapitulation of the instrumental order of capitalism and phallocracy.[3] In Irigaray's terms we are ordered by a Hom(m)osexual "economy of the Same" (Whitford, 1991: 92),[4] where communication and community are limited to exchanges between identical (masculine) subjects. Only "man" is intrinsically valuable. Women and nature are made subject to reason's cold calculations, their reality recognized only insofar as they become hard currency to be valued and traded according to their use.

This materialistic combination of individualism and instrumentalism means that the space for ethical attachments within modernity is both redefined and diminished. Reason dictates that the machinations of modern ethics rely upon the unquestioned presumption of the individual as a moral monad, a solid, bounded, and indivisible entity. Ethics is re-envisaged as a regulatory mechanism for calculating the right and proper apportioning of one's just deserts. Reduced to a series of acontextual rubrics in the hands of professed experts and professional administrators, moral theory becomes a tool for passing judgments on and evaluating the actions of others thereby facilitating managerial and technical efficiency.[5] Modernity replaces Durkheim's mechanical solidarity with Newtons's solid mechanics in a world without either collectivity or conscience, an order without agreed end(s) where individuals are united only by a ceaseless quest for greater efficiency of means.

But there are limits to the explanatory power of such substantialist rationalizations. There are powers that, despite their intangibility, remain omnipotent and omnipresent even in modernity's disenchanted landscape, their existence confirmed by the "scientific" shamans of the concrete and the quantifiable. Newton, the original architect of solid mechanics, had to call upon mysterious forces that act "at a distance" since he could not explain even the falling of an apple without recourse to the suspect and silent activities of that most mysterious of forces—gravity. Even *Homo economicus'* own gurus find the "invisible hand" of market forces behind every action. Yet instrumental rationality's relation to such "insubstantial" powers remains uneasy. For even when named, they still tend to elude our senses. Their presence and power is announced only through their material effects on subjects and objects, in the mysterious movements and motions of matter's constant circulation.

For Irigaray, ethics remains the primal example of such an intangible power. Although ethics can be used by the established order to call the defiant and the deviant into line and, while morality might, on occasion, act as ideological functionary regimenting the unruly masses, its outcomes are always more ambiguous than instrumentality desires. Ethics constantly flies from rationality's iron cage; it refuses to obey the rules of the game. It surfaces in the least expected places with the most unlikely effects; as an instant of selfless companionship, a moment of unsolicited sympathy, or a gesture of defiance that recuperates hope despite a mathematics of overwhelming odds. Ethics is not *sensible*, but its irrationality and intangibility is nonetheless heart-*felt*. Its unexpected surges can sweep us of our feet and its protests against the abuse of authority spring forth even in the most arid of landscapes.

This is why Irigaray describes her project as an *ethics* of sexual difference. She recognizes that ethics is a power that slips by and subverts those who attempt to use it as a technique of domination. Despite his best efforts modernism's *Homo economicus* cannot account for the intransigence and intangibility of the ethical. Instrumental rationality's victory remains incomplete, its hegemony qualified. Beyond the margins and beneath the surface of modernity, overlooked or dismissed, there remains an excess that escapes containment, an undercurrent that flows around and through the bounds of the masculine/instrumental symbolic order: a "physical reality that continues to resist adequate symbolization and/or that signifies the powerlessness of logic to incorporate in its writing the characteristic features of nature" (Irigaray, 1985: 106–7).[6] The modern and the masculine tries to, but cannot, take into account and make tangible that which is least amenable and most alien to its own order of things—the emotional excess that transcends self-interest in its unequal and unfathomable responsiveness to others. The ethical marks a refusal to force others to comply, to make them fit into the categories of one's own symbolic order or to see them as someone exactly like us. In this way Irigaray suggests that we might yet experience a return of this repressed, of the "other('s)" side of the symbolic order; of the feminine rather than masculine, of nature rather than culture, of a mechanics of fluids rather than solids, of ethics rather than economics.[7]

Structuralism, Poststructuralism, and the Subject

I will return to the specifics of Irigaray's conception of ethics shortly. But first I want to elucidate an alternative concept of subjectivity to that s(t)olid, autonomous, and self-interested entity of modernity, *Homo economicus*. With this end in mind I want to examine Louis Althusser's and Judith Butler's accounts of how a relational self might be constituted in terms of her ethical relations to others. I will then turn, in the final chapter, to Pierre Bourdieu's and Henri Levebvre's work to provide the basis of a reflexive theoretical account of the relationship between ethics, subjectivity, and social space.

Modern man seeks, in every instance, to impose a form on others that, while claiming universality, reflects his own instrumental and substantialist concerns. As Val Plumwood has remarked, "[i]nstrumentalism is a way of relating to the world which corresponds to a certain model of selfhood, the selfhood conceived as that of the individual who stands apart from an alien other and denies his own relationship to and dependency upon this other" (Plumwood, 1993: 142). But there are now currents (particularly in feminism and environmentalism) that, in trying to speak for and relate to those others occluded by the current symbolic order, sweep around and against the solidity and immutability of modernist moral discourses. In doing so they challenge the universality of modern man and seek to replace his tedious uniformity with a relational conception of ethics and the self that is sensitive to the vagaries and difference of sex and place (Cooke, 1999). Carol Gilligan's "ethics of care," Judith Butler's analysis of "the psychic life of power," and Luce Irigaray's "ethics of sexual difference" each, in their own way, emphasize ethics as *excess*, as that which refuses to be contained within the confines of the modernist and masculine order. They posit an ethical self that is relational, ambiguous, and intangible, an alternate reality that continues to resist instrumentalism's attempts to control women and wild(er)ness.

Carol Gilligan's work has been of immense import in challenging the dominant masculinist moral order. As chapter 6 indicated, she "calls into question the values placed on *detachment* and *separation* in [those] developmental theories" (Gilligan, 1994: 213) like the ontogenetic model proposed by psychologist Lawrence Kohlberg (Heckman, 1995; Larabee, 1993). Where Kohlberg equates ethical maturity with attaining the ability to remain "objective," *to stand back* and apply abstract rules and principles to moral problems, Gilligan emphasizes the ethical subject's *closeness* and *attachment* to others. Her moral agent is not an egocentric automaton but a relational self, that is, a person whose relations to others are constitutive of rather than incidental to her identity. Rather than developing a moral theory with universalistic pretensions, Gilligan advocates paying attention to the differing situations and contexts in which we actually concern ourselves with others. This ethics of care regards the ethical as a vital (i.e., living) dimension of self-identity that is not reducible to subjective self-interest nor is it simply a consequence of being forced to accept the self-effacing mechanical reproduction of objective and external social laws.

This latter point is important. Modernity's *Homo economicus* tends to regard all "external" influences as infections that threaten his integrity but ethics need not entail self-abnegation. This is not to say that ethics is blameless in compromising the hermetic isolationism of the individual, or that it is somehow innocent of any involvement with our confinement within the current symbolic order: quite the contrary. Those social theorists (like Durkheim or Althusser) who have emphasised ethics' function in inculcating social norms and imposing order on nascent individuals are not entirely wrong. We are not free spirits who live in an ideological vacuum but beings whose identities actually coalesce within the fields of gravity that pervade the spaces of our social milieux. We are constituted, at least in part, through our immersion in the dominant symbolic order and, in becoming subjects, we also incorporate as our own the grammar and syntax of the prevailing cultural system, its language, its values, and so on.

This relational view of subjectivity and social reality is common to all of those theorists that employ a spatial metaphorics in order to foreground the relationship between the symbolic order and subjectivity. Such theorists obviously have close links with structuralist and poststructuralist philosophies since these have always employed spatial metaphors (Silber, 1995). For example, both Althusser's and Bourdieu's account of social space seems, in many ways, analogous to Saussure's description of the structure of a language. For Saussure, each word gets its meaning, not by possessing an inherent essence that fixes its signification in relation to that it signifies but because of its place—its role—in relation to all the other words that compose that language. Saussure puts it thus—"signs function . . . not through their intrinsic value but through their relative position" to other signs (Saussure in Kirby, 1997: 29). Words are never autonomous, rather their "content is really fixed only by the concurrence of everything that exists outside it" (Saussure in Kirby, 1997: 28), that is, the meaning of any one word is determined by the system of language as a whole. Initially then we might be tempted to think that the manner in which Saussure sees one word as distinct yet dependent upon another is similar to the way that the subject's self-identity is delineated through its relations to others.

This is precisely the insight behind Althusser's structuralism. As chapter 3 explained, Althusser believed that all social formations are dynamic but structured systems composed of relatively autonomous and nonreducible spheres, that is, the economic, the political, and the ideological spheres. Althusser counters tendencies within Marxism toward economic reductionism, arguing that "the superstructure [the political and ideological] is not the pure phenomena of the structure [the econonomic], it is also the condition of its existence" (Althusser, 1969: 205). Althusser's antireductivism leads him to emphasize the irreducibility and distinctiveness of each part of the system. But this structural approach is also relational in that it also refuses to look for the essence of things-in-themselves. The economy is not regarded as fixed and foundational but only as a *relatively* autonomous level within the social structure. This is because the economic structure is not envisaged as a fixed framework upon which society hangs. We cannot look beneath the skin of society and expect to find the structure labeled as if it were a feature on an anatomical atlas. This structure is not like the "skeleton" or "musculature" of a body. Rather the structure simply is the manner in which the "parts" of the social formation are constituted with respect to each other and the "whole." The structure *is* the dynamic and determining interrelations between the economy, the political, and the ideological. It is not a separable pre-existing framework or an original cause, but an immanent expression of the articulation of social practices with each other.[8] For Althusser, the real is relational.

What is true for society is no less the case for the individual subject. Althusser's structural antihumanism refuses to recognize the existence of a human essence, a shared "human nature" buried deep within all of us. This transcendental subject who would remain the same in all times and places is, for Althusser, a humanist myth entailing the reification of a particular conception of humanity. While the humanist claims to elucidate an *a priori* and universal notion of what it is to be human, they are actually engaged in making *a posteriori* generalizations derived from a particular and historically embedded idea of the subject. (This is why humanists cannot agree among themselves on a model of human nature.) As Althusser makes plain, this search for a human essence also depends upon precisely the kind of substantialism that any relational perspective abhors, for, "[if] the essence of man [sic] is to be a universal attribute, it is essential that *concrete subjects* exist as absolute givens; this," Althusser claims, "implies an *empiricism of the subject*" (Althusser, 1969: 228). That is, humanism is a form of substantialism that in seeking the grounds of *solidarity* between concrete individuals tries to identify a core of shared characteristics. Yet, to serve its purpose, this core cannot be the property of any one individual alone but must be shared by all. Thus ironically, modern humanism falls back into just the kind of mysticism it wants to repudiate, that is, it invents a mysterious and insubstantial human essence analogous to the invisible hand of the market. Humanism's reliance on the empirical obviousness of the individual as its grounds for solidarity (its subjectivism) is really just the other side of an (objectivist) coin. Neither position is capable of understanding the constitutive relations between self and society.

For Althusser, Marx's revolutionary step was precisely to reject the transcendental humanism that had previously been common to all post-Enlightenment (bourgeois) conceptions of the subject, whether political (from Hobbes to Rousseau), ethical (from Descartes to Kant), or economical (from Petty to Ricardo) (Althusser, 1969: 228). The mature Marx recognized that the nature of individual subjectivity differed markedly from one social formation to another. Thus, Althusser seeks to replace the humanist question of the "role of the individual in history" with one of determining the "historical forms of existence of individuality . . . the different forms of individuality required and produced by that mode [of production]" (Althusser & Balibar, 1986: 11).

In this way the subject is relativized, each individual is differentially produced through their inclusion at a particular location within a particular social formation. Becoming a subject is a process of self-recognition that entails that the individual comes to recognize their place, their position and role, within that society. This psychopolitical process, which

Althusser refers to as "interpellation," is primarily the result of the internalization of the prevailing ideology during the nascent subject's formative years. Indeed this is the primary role of ideology, which "[i]nterpellates [i]ndividuals as [s]ubjects" (Althusser, 1993: 44).

It is crucial to understand that Althusser's conception of ideology is *not* one of false consciousness. Ideology is neither false nor, for the most part, conscious. Ideology is an absolutely necessary part of any society, even communist societies, since all societies survive and reproduce themselves via the inculcation of the particular kinds of behavior, values, and so on that subjects carry into the next generation.[9] We are all subjects, and live in ideology. "Man [sic] is an ideological animal by nature" (Althusser, 1993: 45). Nor should the interpellation of the individual in ideology be confused with a person coming to accept a particular political creed or a narrowly defined worldview. Ideology operates at an altogether different level from this—in structuring the configuration of a person's relations to the world. It is not as given *ideas* that ideology functions, but as structures delimiting the possible frameworks in which any thought at all takes place. "[I]n the majority of cases these representations have nothing to do with 'consciousness'; they are usually images and occasionally concepts, but it is above all as structures that they impose upon the vast majority of men. [sic]" (Althusser, 1969: 233).

How then does ideology operate to bring about the interpellation of the subject? For Althusser, language as the key conveyor of the symbolic order plays a crucial role in the process of interpellation.[10] The nascent subject finds their identity by insinuating themselves, and being insinuated, into the pre-existing symbolic order; they gradually recognize themselves as bearing a certain relationship the structures of a ready-made language, codes, and so on. Thus, in Athusser's most famous example, the subject hears the policeman's shout "Hey, you there" and in turning round implicitly accepts that it is addressed to him, that is, he recognizes himself, qua subject, as the one who is being called to task (Althusser, 1993: 48–49). This then is what Althusser means when he states that "ideology *hails* or interpellates concrete individuals as concrete subjects" (Althusser, 1993: 47, our emphasis).

However, one shouldn't misconstrue ideology's apparent effectiveness in calling the subject into being as somehow due to its inculcation within the subject of an accurate representation of society and her subject position. The symbolic order carries within it the imbalances and distortions of the society it claims to represent. And, as a Marxist, Althusser recognizes that ideology serves the interests of the ruling classes. To this extent the subject's recognition of their position is (in capitalism at least) inevitably a *misrecognition* (*méconnaissance*) since, "what is represented in ideology is . . . not the system of real relations which govern the existence of individuals, but the imaginary relation of the individual to the real relations in which they live" (Althusser, 1993: 39).[11] "[I]t is the *imaginary nature of this relation* which underlies all the imaginary distortion which we can observe (if we do not live in its truth) in all ideology" (Althusser, 1993: 38).

The Subject of Conscience

Althusser was by no means alone in announcing the "death of Man," that transhistorical hero of Enlightenment philosophy who, it was supposed, was to be found in us all. (Or at least in all men.[12]) Structuralists and poststructuralists in anthropology (Lévi-Strauss), philosophy (Foucault), psychology (Lacan), literature (Barthes, Goldmann), and so on set themselves explicitly against what they regarded as a concept that both reified the human essence via an unwarranted metaphysics and, at the same time, reduced our ability to comprehend human life to the banal empirical investigation of concrete individual "billiard balls." This motif also, of course, becomes one of the main characteristics of those theories later referred to as "postmodern."[13] On the other hand, humanists, including humanist Marxists like E. P. Thompson, launched vitriolic attacks on a theory they saw as undermining the very rationale for, and possibility of, radical social change (Thompson, 1978; Soper, 1986). In whose name was the revolution to be fought if not that of humanity? If our view of ourselves as empowered beings with free will is only due to ideological misrecognition, what sense can we make of the idea of freedom at all?

In removing the individual as a subject of history Althusser seems to leave little or no room for either individual agency or for the critique of ideology's pervasive effects. His account of the relation between subject and society apparently privileges social structures over the subject's agency. Indeed, as many have remarked, Althusser's structuralism comes close to a reductive functionalism where individuals have no import except insofar as they fulfill the roles allocated to them by society. Subjects are simply nodes in a social nexus over which they can exert no control. "[T]he structure of the relations of production determines the places and functions occupied and adopted by the agents of production, who are *never anything more than the occupants of these places*, insofar as they are the "supports" (*Träger*) of these functions" (Althusser & Balibar, 1986: 180, our emphasis).[14] This structuralist conception of the human subject as a bearer of social roles is in direct opposition to the humanist vision of the autonomous rational agent, a difference that marks the poles of the structure/agency debate in social theory.[15] The words "never anything more than" certainly imply that individuals per se count for little, that they are passive social products wholly determined by forces beyond their control. Althusser's structuralism does little to dispel this notion for, despite his recognition that ideology is not logically prior to subjects, but is itself constituted by the practices of individual subjects (i.e., that the twin realities of ideology and individual identity are relational), there seems little doubt that there is only a limited reciprocity in this constitution. "*[T]he category of the subject is only constitutive of all ideology, insofar as, all ideology has the function (which defines it) of constituting concrete individuals as subjects*" (Althusser, 1993: 45).

There are, then, two related issues that Althusser's problematic fails to fully address. Both stem from the fact that his relational account of the subject actually comes close to entailing the dissolution of that subject. First, his antihumanism seems to imply that there is no subject, no concrete individual prior to their interpellation in society. But this is surely paradoxical for, as Judith Butler notes, "the narration of how the subject is constituted presupposes that the constitution has already taken place" (Butler, 1997: 11). If not, then *who* is it that the policeman hails? In other words, there is a need to posit a "site" where "subjection" takes place and where can this be if not in a concrete, substantial individual, a subject who already shares with others certain properties—even if these properties merely take the form of predispositions to hear and recognize the voice that hails them? Such considerations might seem to return us to a variety of humanism. Second, the emphasis on the subject being "nothing more than" a support of the system overemphasizes the subject's subordination to the prevailing ideology, removing all that subject's power of agency and of self-constitution. The subject thereby seems to lose the "relative autonomy" that is usually taken to be its defining feature.

The Althusserian subject seems therefore to be a somewhat paradoxical entity. Yet this need not prove fatal, since subjectivity is indeed, as Judith Butler points out, necessarily ambiguous. We come to recognize ourselves *qua* subjects through the operation of those very powers that constitute us. The process of interpellation or subjection (*assujetisement*) implies that in becoming a subject one also becomes *subject to* these interests. In Butler's words, "'Subjection' signifies the process of becoming subordinated by power as well as the process of becoming a subject . . . power that at first appears as external, pressed upon the subject, pressing the subject into subordination, assumes a psychic form that constitutes the subject's self-identity" (Butler, 1997: 2–3). This recognition of our self-identity comes in the allegorical moment of the turn towards the voice of authority. The very process of self-recognition is, at one and the same time, a turning of the subject against itself as we "embrace the very form of power—regulation, prohibition, suppression—that threatens one with dissolution" (Butler, 1997: 9). Such "[p]ower is not simply what we oppose but also, in a strong sense, what we depend upon for our existence and what we harbor and preserve in the beings that we are" (Butler, 1997: 2).

The subject-to-be is necessarily vulnerable when born,"[b]ound to seek recognition of its own existence in categories, terms, and names that are not of its own making" (Butler, 1997: 20). To become a subject then it seems one must have some primary "longings, desires or drives" that allow, indeed compel one to assume a certain subject position, to "find oneself" (in both senses) within the social and symbolic order. In this sense the nascent subject is complicit in his or her own subjection, that is, subordination. But this complicity can itself only be ensured through the subject's *self-regulation*, through the turning back of these "drives" onto themselves in such a way as to challenge the emergent subject about the degree and nature of their compliance with prevailing social norms and thereby prohibit or foreclose certain paths of "abnormal" development. In other words "[t]he doctrine of interpellation appears to presuppose a prior and unelaborated doctrine of conscience, a turning back upon oneself" (Butler, 1997: 109). *Conscience* is the first sign of agency (of self-consciousness) and partly constitutive of agency; it marks the reflexivity that defines the psychic space of the emergent self. Ethics precedes ontology, since "[c]onscience is the means by which a subject becomes an object for itself, reflecting on itself, establishing itself as reflective and reflexive. . . . In order to curb desire, one makes of oneself an object for reflection" (Butler, 1997: 22).[16]

In Butler's words, the "turn toward the law is . . . a turn against oneself, a turning back on oneself that constitutes the movement of conscience" (Butler, 1997: 107). And yet this complicity also installs the potential for subverting the social order because the one addressed can then begin to ask "a set of critical questions: Who is speaking? Why should I turn around? Why should I accept the terms by which I am hailed" (Butler, 1997: 108). We may be constituted within an ideology and a social formation over which we have little or no control but the very act of our constitution opens up the possibility of our turning "against" that power: We might "resist"/"recuperate" it for our own purposes. We are enabled as subjects because the "power that initiates the subject fails to remain continuous with the power that is the subject's agency" (Butler, 1997: 12). Agency does not remain tied to the conditions of its production.

Accepting the broad outline of Althusser's problematic does not then necessarily have to mean that we lose sight of the individual subject. We are not reducible to *Träger* of the social formation. (Indeed, Althusser himself recognized that he may have gone too far in his critique of the philosophical humanism that seemed to pervade Marxist circles following the publication of Marx's early "Hegelian" works. "I did consciously think in extremes about some points which I considered important and bend the stick in the opposite direction" [Althusser, 1990]).

"Agency [then] exceeds the power by which it is enabled" (Butler, 1997: 15). But in his haste to proclaim his in-dependence from others, modern man seeks to exaggerate and fix this discontinuity between self and other. Hoping to quell the amniotic chaos of his birth by acquiring hard and fast boundaries on his identity he strikes out for dry land and its sureties. He severs all links to others and, forgetting the aqueous ambivalence of his origins, would have us believe that he was always thus, an autonomous agent; a self-made man. The logic of this masculine/modernist symbolic order refuses to accept our symbiotic status, but as Butler points out, "exceeding is not escaping" (Butler, 1997: 15). The "subject exceeds the logic of non-contradiction" only insofar as it is "*neither* fully determined by power nor fully determining of power (but significantly and partially both)" (Butler, 1997: 17). Butler's psychological exegesis of Althusser does not then presage a return to the humanist subject precisely because the subject per se does not exist prior to the process of subjection. The subject "is an excrescence of logic, as it were" (Butler, 1997: 17), an excess that confounds the phallocratic order of things but is never wholly independent. From this perspective the ethical is not an alien infringement on the autonomy of a separable and pregiven self but a constituent and primary *passion*, an ineradicable ambiguity at the *heart* of a relational self.

Here then is the true irony of interpellation since the subordination of the subject is guaranteed via internalization of social norms in such a way that the subject must come to regulate their own attachment to those norms. But this very act of self-regulation inevitably requires a self-recognition that entails the instauration of a potential for difference as well as compliance. Far from being a mechanical guarantor of social conformity the "ethical" is inherently ambivalent. Not only is it the conservative instrument of social reproduction it is also the primary source of difference, the very opening through which dissent can flow into the social realm. Any attempt made to channel and control our ethical activities is never entirely successful since "fluid is always in a relation of excess or lack vis-à-vis unity. It eludes the "'Thou art that.' That is, any definite identification" (Irigaray, 1985: 117).

Desire and Wonder: On *Being* Touched by the Other

In a way, wonder and desire remain the spaces of freedom between the subject and the world.
—Irigaray, *An Ethics of Sexual Difference*

We tend to think of power as an external force that "presses upon the subject from the outside, as what subordinates, sets underneath and relegates to a lower order" (Butler, 1997: 2). But, as Butler argues, things are rarely so simple. All power is ambiguous and as Avery F. Gordon remarks, power relations are "never as transparently clear as the names we give to them imply. Power can be invisible, it can be fantastic, it can be dull and routine. It can be obvious, it can reach you by the baton of the police, it can speak the language of your thoughts and desires. It can feel like remote control, it can exhilarate like liberation, it can travel through time, and it can drown you in the present. It is dense and superficial, it can cause bodily injury, and it can harm you without seeming ever to touch you. It is systematic and it is particularistic, and it is often both at the same time" (Gordon, 1997: 3).[17]

The ambiguity inherent in power's application is nowhere more pronounced than in the process of our subjection, a process that installs within us a conscience, a "passionate attachment" (Butler, 1997: 7) to others. This "primary passion," a desire initially born of a state of dependency, can only fulfil its mimetic function by giving rise to self-consciousness, through the recognition of ourselves as different from those we would emulate. In this sense, self and other are co-respondent, the one requires the other but is not reducible to her. In Irigaray's words, we "look at the other, stop to look at him or her, ask ourselves, come close to ourselves through questioning. *Who art thou? I am* and *I become* thanks to this question" (Irigaray, 1993: 74). The ethical relation is "a chiasmus or a double loop in which each can go toward the other and come back to itself" (Irigaray, 1993: 9).

But modernity/masculinity seeks to deny the intangible presence of others in our hearts, to reinforce the concrete and impervious boundaries of an autonomous self as an objective and easily grasped entity. Modern man's insistence that difference be accounted for according to his own phallologocentric rationale reduces all others to imperfect copies of himself. Whether the denial of ethics' dialectic takes the form of the sublimation of the self or the repression of the other, the end result is—the Same. (That is, the ethical is eradicated in favor of a monopolar, hom(m)osexual economy of the Same.) Should conscience become mere compliance, should woman and nature become consumed by man's self-love, by his desire to eradicate all difference, then the space for the ethical is diminished, flattened to one dimension.

Irigaray realizes, where Butler does not, that desire is not enough to ensure the existence of the ethical. "Desire occupies or designates the place of the interval. . . . Desire demands a sense of attraction: a change in the interval, the displacement of the subject or of the object in their relations of nearness or distance" (Irigaray, 1993: 8). But unalloyed attraction will eventually eradicate difference. What Butler never explains is why the emergent self should ever choose to question its own attraction to the other it seeks to emulate, or how it can be *moved* by the other's difference while being pulled within the circle of desire's gravity. Why, even given an emergent self-consciousness, doesn't the subject proceed to consum(at)e herself in a mimetic frenzy? Why are we not under a compulsion to comply in every way with the dominant symbolic order? In other words, Butler does not really explain how an ethical relation that respects the differences between self and others is possible or how an ethics of sexual difference might resist the economy of the Same.

From Irigaray's perspective "[w]hat is missing is the double pole of attraction and support, which excludes disintegration or rejection, attraction and decomposition, but which instead ensures the separation that articulates every encounter and makes possible speech, promises, alliances" (Irigaray, 1993: 9). If desire is an attractive force, the "silent gravity" of love, then wonder is the elemental passion that can compose a space for difference, the "thin air" that separates and allows us to keep a respectful distance. Wonder is a "sudden surprise of the soul which causes it to apply itself to consider with attention the objects which seem to it rare and extraordinary" (Descartes in Irigaray, 1993: 77).[18] We stand amazed, our bodies and brains infused by that "movement of . . . spirits" (Descartes in Irigaray, 1993: 77) within us occasioned by others whose intangible difference lies beyond our grasp. "[W]onder goes beyond that which is or is not suitable for us. The other never suits us simply. We would in some way have reduced the other to ourselves if he or she should suit us completely. An *excess* resists: the other's existence and becoming as a place that permits union and/through resistance to assimilation or reduction to sameness" (Irigaray, 1993: 74). In wonder the other "should *surprise* us again and again, appear to us as *new, very different* from what we knew" (Irigaray, 1993: 74).

Wonder as a "passion has no opposite or contradiction and exists always as though for the first time" (Irigaray, 1993: 12).[19] And so reason fails to comprehend that which, in the uniqueness of its manifestation, exceeds phallocracy's repetitive logic. Such feelings are alien to the modern man, their very immateriality an offence to his sensibilities. But "wonder is an *opening*" (Irigaray, 1993: 81), a passion that ruptures the rational order in its quest to know the other in all her difference, "an attraction and return all around the unexplored, all barriers down, beyond every coast, every port. Navigating at the center of infinity, weightless" (Irigaray, 1993: 80).

Wonder tempers desire's attractive force, engenders respect for the space between us. Where desire alone would possess the other, define her and occupy her, "[t]he 'object' of wonder or attraction remain[s] impossible to delimit, im-pose, identify (which is not to say lacking identity or borders): the atmosphere, the sky, the sea, the sun" (Irigaray, 1993: 81). The ethical subject thus emerges through a process that inaugurates rather than eradicates her difference to others. Even if our original desire is to conform we are struck with wonder at the panoply of difference. Modern "man" cannot resist this power since "[w]onder is a mourning for the self as an autarchic entity; whether this mourning is triumphant or melancholy. Wonder must be the advent or the event of the other. The beginning of a new story?" (Irigaray, 1993: 75).

Ethics is then the flow of *things* in desire and wonder, it is a relation that lets things *be*, conserving and sustaining them in love and/of difference. Rather than isolating its object in an economy of the Same, ethics tentatively embraces, but never *envelops* the other in a relation that preserves difference.[20] The ethical relation is never a Newtonian mechanics of solids, of equal and opposite cause and effect, it is an asymmetrical and nonreciprocal relation of excess. It asks nothing of the other but is touched by the other's presence in itself.[21] The ethical subject is never "identical" with any other being, she is unique but never alone, an eddy in the ebb and flow of a relational matrix of difference, a matrix that is, primarily, ethical.[22]

Such a relation falls outside of modernity/masculinity's comprehension since this dominant symbolic and cultural order gives primacy to ontology, to the tangibility of *things*, promoting a solid mechanics that operates according the tyranny of reason's (instrumental) principles. But as Irigaray makes clear in her rereading of Emmanuel Levinas, despite its intangibility, the ethical is a "relationship with the Other as interlocutor [that] precedes all ontology; it is the ultimate relation in Being" (Levinas, 1991: 48). The ethical is (like the feminine) "*what is in excess with respect to form* [. . . that which] *is necessarily rejected as beneath or beyond the system currently in force*" (Irigaray, 1985: 110–11). Ethics is a "never-surfeited sea" (Shakespeare, 1971: 59) that tends toward *infinity* rather than *totality*, it cannot be fixed, frozen, or fully accounted for.

The tangibility of the ethical is of a different kind, an intimacy, a proximity, that is at once distinct from the exclusivity of erotic intimacy but nonetheless passionate. It is both corporeal and "spiritual," physical and metaphysical, an "encounter between the most material and the most metaphysical. . . . Neither one nor the other. Which is not to say neutral or neuter" (Irigaray, 1993: 82). Its touch is light but profound, respectful and desirous of changing us in our innermost being. Rather than imposing on the other, manipulating and molding them for our own uses its touch will "place a limit on the re-absorption of the other in the same. Giving the other her contours, calling her to them . . . inviting her to live where she is without becoming other, without appropriating herself" (Irigaray, 1993: 204).

Ethics entails recognizing that we cannot occupy another's place. It is giving the other the space to be as s/he will without any expectation of recompense. Its anarchic motion follows no laws or principles, its currents refuse to be constrained, its oceanic depths cannot be fathomed. The ethical subject emerges in a relation that inscribes difference as well as similarity at her very heart.[23]

The Ecological Self Revisited

Irigaray's conception of ethics as excess indicates how a relational subject, who is beholden to a prevailing ideology for her very existence, might still transcend the logic of her circumstances in her ethical attachment to and care for others. Ethics eludes the limitations of the prevailing modern/masculine symbolic order because it resists its tendency to reduce all others to an economy of the Same. The ethical relation to the other is one of desire *and* wonder, of empathy *and* amazement. It recognizes but does not seek to diminish the other's difference from ourselves, letting them be(come) what they will. Of course, for Irigaray this difference is primarily sexual. Women must struggle to express themselves in a linguistic and social order constructed by and for men. For this reason Irigaray's feminism eschews any attempt to achieve equality within a social syntax that is always already masculine. The egalitarian approach accepts the arbitrary phallocratic assumptions underlying the present symbolic order and restricts women to an economy of the Same. But to be themselves women must be able to create a language and space that can adequately express their difference. This entails the development of a new *ethical* relation between the sexes. Irigaray argues that, should this occur then "sexual difference would constitute the horizon of worlds more fecund than any known to date" (Irigaray, 1993: 5).

From the perspective of a radical environmentalism Irigaray's work has many advantages. It provides a ready critique of that instrumentalism that, in Val Plumwood's words, "is a mode of use which does not respect the other's independence or fullness of being . . . and strives to deny or negate that other as a limit on the self and as a center of resistance" (Plumwood, 1993: 142). Irigaray (like Butler) also undermines that hyperseparated and self-interested subject whose "ends have no non-eliminable reference to or overlap with the welfare or desire of others" (Plumwood, 1993: 144). Hers is both a work of anamnesis, a remembering of the intimate and irrevocable connections to others that constitute every individual, and a call for patterns of difference and diversity to be *conserved* and *sustained*.

However, Irigaray's emphasis on sexual difference to the exclusion of all other aspects of difference (including class, race, etc.) also has the obvious disadvantage that the significant other is reconfigured in almost entirely anthropocentric terms; She/he is always human. Despite her frequent use of environmental metaphors, of sea and sky, air and water, and her to allusions to the female body, nature usually only enters Irigaray's ethical discourse in human form. Irigaray's relational self is not then an "ecological self" that might, in Plumwood's words, include "the flourishing of earth others and the earth community among its own primary ends, and hence respects and cares for these others for their own sake" (Plumwood, 1993: 154). Nonetheless, there is nothing intrinsic to Irigaray's work that speaks against extending it in this environmental way and to do so might help avoid a tendency among some advocates of an ecological self to "plunge us into the problems of merger and the denial of difference" (Plumwood, 1993: 155). That is, Irigaray might help us avoid the tendency in contemporary ethical theory to recognize our moral relationship to nonhuman others in either one of two ways. First, by erecting a scale of similarity, an anthropocentric hierarchy, that regards nonhumans as more or less similar to humans and values them accordingly. Second, by abandoning all distinctions and risking falling into a holistic or monistic morass that threatens to be every bit as corrosive to the subject's identity and agency as Althusser's structuralism. As chapter 1 argued, neither way gives due recognition to the other for what they are in themselves. Both reduce nature to that economy of the Same that is the modernist/masculine symbolic order.

From an ecological point of view it is understandable that environmentalists should try to combat modernism's hom(me)ocentric instrumentalization of nature by either making nature an honorary human or by emphasizing the spiritual possibilities of sublimating the human self within nature. Both aspects are prominent in the writings of key figures in the conservation movement like John Muir, the expatriate Scotsman who played such a pivotal role in founding the Sierra Club and in campaigning for national parks in the United States. Muir, the doyen of deep ecology, describes the view from Twenty Hill Hollow, a valley near his beloved Yosemite: "To lovers of the wild, these mountains are not a hundred miles away. Their spiritual power and the goodness of the sky make them near, as a circle of friends. . . . You bathe in their spirit-beams, turning round and round, as if warming at a camp-fire. *Presently you lose consciousness of your own separate existence: you blend with the landscape, and become part and parcel of nature*" (Muir in Oelschlaeger: 1991: 188). Despite their geographical distance the mountains eventually seem so close to us that we lose all sense of separation from them. We are as one.[24]

No doubt this kind of identification with nature strikes a chord with many of us precisely because there is so little opportunity to experience nature's otherness in the modern world. We hope to find in nature an alternative to the alienating order of everyday urban existence. Muir's pantheism recognizes a sacred aspect to places like Yosemite whose grandeur transcends the profanities of modernity; he speaks of his "resurrection day" and his "baptism" in these natural places (Muir in Oelschlaeger, 1991: 187). But, despite Muir's religiously inspired antimodernism and his dedication to wilderness conservation, his account of nature may be unsuitable for an environmental *ethics*. Muir's indubitable sense of awe and wonder is all too often overcome by the desire to eliminate difference by making nature almost homely or surrendering himself to its overpowering presence.[25] In an ironic echo of Müller's naturalism (see chapter 2), he lies "humbly prostrate before the vast display of God's power, and eager to offer self-denial and renunciation with eternal toil to learn any lesson in the divine manuscript" (Muir, 1988: 79).

Muir sees the wilderness as a timeless source of spiritual regeneration where we can lose our modern cares and unburden ourselves of uncertainties. "No pain here, no dull empty hours, no fear of the past, no fear of the future" (Muir, 1988: 79). The wilderness not the city is humanity's natural home, we merely need to attune our senses to it, to read it anew as a source of comfort and enlightenment. But while nature may be the "cradle" of humanity, Muir's somewhat bowdlerized tales abound with anthropic reference more suitable to the Victorian nursery. The overhanging cliff is there to shelter us from the mountains storms, ferns and mosses are "reassuring . . . gentle love tokens," daisies are "children of light, too small to fear. To these one's heart goes home, and the voices of the storm become gentle" (Muir, 1988: 80). Without taking away from Muir's immense achievements we must question whether either moralistic homilies or pantheistic self-abandonment will in the long term help or hinder the autopoeitic flow of nature's difference. Perhaps the motorized multitudes who now descend on Yosemite threatening to "love it to death" really are Muir's philosophical progeny.

By contrast, an Irigarayan environmental ethics would mark that excess that refused at all costs to assimilate nature's other to the Same. "An *excess* resists: the other's existence and becoming as a place that permits union and/through resistance to assimilation or reduction to sameness" (Irigaray, 1993: 74). An ethical relation to nature cannot be expressed by either an anthropic erasure of difference or by the sublimation of the self in ecstatic raptures. An environmental ethics must let nature be (verb) allowing it to manifest itself within life and language in all its uncomfortable (in)difference. The difficulty of the ethical relation lies precisely in maintaining a creative tension between expressing care, consideration, and even closeness while simultaneously conserving space for the other's being so that it remains uncompromised by our presence.

This is not to criticize Muir's use of religious terminology in his, often illuminating, descriptions of his wilderness experiences, but an ethical relation requires that we renounce inappropriate iconography and avoid falling into the selfless worship of the devotee. Interestingly, Max Oelschlaeger describes Muir's experience more appropriately as an epiphany, a term also used by Levinas. Ethics is, Levinas argues, an epiphany whereby the other is revealed in the fullness of its difference to us. We experience a realization, a moment with which the other is made real or manifest to us and where we are brought to recognize its ethical standing in the play of desire and wonder. The other is not thereby *reduced to us* but is *revealed* as something that must be conserved in its difference to us.

This epiphany can happen in innumerable and unpredictable ways. *There are no rules where ethics is concerned.* Like love this relation can take years to unfold, gradually settling in the core of our being. Alternatively, it can appear in an instant, a lightning bolt that strikes to our heart revealing a world we had never imagined existed, a world where things near and far are thrown into sudden relief against the previously existing gray sameness. One could watch documentary after documentary on the plight of whale or rainforest and yet never really be touched. And then, a single instant or image, the sound of a dolphin's screams in the tuna catcher's nets, a chainsaw ripping through an oft-passed tree, or alternatively the soft touch of a snowflake falling or the Sequoia's cushioned bark lets the *real* world rush in on us. Our preconceptions and complacency are swept aside in the streaming of the ethical "moment"—the movement of "excessive" force—that sweeps us up and sets us down in a place that is never to be the Same—facing an altered understanding of self and situation.[26]

This revelation can often prove far from comfortable. Looking into the face of the other subverts our self-assuredness and we are challenged to place ourselves in jeopardy. But such discomfort cannot be resolved by sublimating the self, by attempting to become one with the other, any more than they can be resolved by attempting to make the other one with myself. Rather, the ethical entails maintaining a critical relation, it requires a critical conscience that both bears the other in mind and puts oneself into question. "The putting into question of the self is precisely a welcome to the absolutely other" (Levinas in Peperzak, Critchly & Bernasconi, 1996: 17).

But if the ethical *momentum* both problematizes the self and conserves the play of difference between self and other, then the fixed and determinate logic of the moral law and modernist moral philosophy is an ethical vacuum that it abhors. Both nomological and axiological interpretations of ethics attempt to petrify the flow of ethical feeling, to regulate and contain difference in a repetitive acontextual logic that requires and exemplifies the repetition of the Same. They seek to delimit boundaries and deny the very existence of relations that *exceed* or contradict their formal rules. Such logic presupposes relations of equivalence and exclusivity, of identity and nonidentity, but ethics glories in confounding such clear-cut distinctions and regularities. "The other does not show itself to the I as a theme" (Levinas in Peperzak, Critchly, & Bernasconi, 1996: 17) but as a passion.

Levinas, like Irigaray, rejects attempts to regulate and subdue the ethical moment, to replace love with logic and desire with duty within modernity's solid mechanics. Such attempts to impose a narrative of totality fail to grasp ethics' intangible and relational excess.[27] Ethics is anarchic—it refuses to be made fixed and determinate, to be subject to the stamp of authority. Levinas continues, "the putting into question of the Same by the Other is a summons to respond. The I is not simply conscious of this necessity to respond, as if it were a matter of an obligation or duty . . . rather the I is, by its *very position*, responsibility through and through" (Levinas in Peperzak, Critchly, & Bernasconi, 1996: 17).

Whether made manifest as the passionate product of a fleeting moment or the quiet surfacing of long-present but hitherto unsuspected depths, ethics remains intangible. Yet this does not mean that it does not leave its mark on self and other. Like the river's flowing, the ethical stream shapes the landscape of our lives, it springs forth from hard rock, cuts deep canyons, carries and deposits the alluvium of our hopes and dreams. Its ripples echo through the soul.

8

A Green Thought in a Green Shade: Moral Sense and an Ethics of Place

Recall yourself once more, I insist, into the air.
—Irigaray, *Elemental Passions*

In this, the final chapter, I want to pursue the question of the "ethical subject" further by extending and elaborating a theory of the relationship between the "phenomenology of ethical feeling" and the "production of ethical spaces." This will allow me to trace a conception of ethics as an *excess* that can resist the rationalization and the disenchantment of the life-world, a power that can sustain that anarchic love of nature that is a characteristic feature of radical ecology.

As the last chapter illustrated, Irigaray deploys a counterhegemonic discourse that recognizes ethics as a primary passion, a flow of desire and wonder that is constitutive of subjectivity. Ethics, so conceived, is a site of resistance that in its openness to difference eludes all attempts to make it tangible and manipulable and thereby reduce our relations with others to an economy of the Same. But if we can't *grasp* ethics, if we can't fathom its reasons, capture its essence, or delimit its scope, as ethical axiologies have striven to do, then how are we comprehend it at all? Butler and Irigaray both approach this question from psychosocial perspectives that are ideally suited for tracing the early origins of subjectivity but perhaps fail to give a ready account of the relatively persistent ethical relations between self and significant others in wider social and natural circumstances. In other words we need to understand how our ethical values come to *inform* our everyday lives, to characterize our patterns of behavior and those lasting relationships between persons and places that are renewed constantly yet can frequently survive even a prolonged separation.

This reconception of ethics clearly cannot be couched in the same terms as the formal ethical theories it seeks to replace. Instead, it must reveal and relate a mode of ethical existence that explicit systems of rules, principles, and calculations, like rights based and utilitarian approaches, have ignored or subsumed. Instead of employing modernity's acontextual and disembodied rationality to order and direct our relations to others, this reconception of ethical being must speak of the constitutive relationship between an embodied ethical experience, that might only rarely operate at the level of full consciousness, and specific contexts, that is, environments. In this way we might develop a genuine "ethics of place."

The social theory of Pierre Bourdieu is relevant here since it concerns itself with the construction of subjectivity in social space in order to illuminate the manner in which norms are both expressed and subverted in individual actions. Bourdieu's work, like Butler's, arose as a direct response to his dissatisfaction with those varieties of structuralism, like Althusser's, that, by emphasizing the import of "external" relations on the constitution of the self, had seemed to many to remove all possibility of agency and of self-constitution.[1] "I wanted, so to speak, to reintroduce agents that Lévi-Strauss and the structuralists, among others Althusser, tended to abolish, making them into simple epiphenomena of structure. And I mean agents, not subjects" (Bourdieu, 1990: 9).[2] Bourdieu is highly critical of the determinism implied by Althusserian models but never falls back into an atomistic individualism in his attempts to resolve this structure/agency debate.

Because of the prevalence of spatial metaphors, Bourdieu's work might be regarded as a social *topology* of the subject. For him, one's identity, qua individual subject, is constituted in and recognized through the position one occupies in relation to significant others. One's identity is not an essential internal attribute that one carries unchanged through life, but arises from the complex interrelationship between self and society where the individual's uniqueness is itself a function of standing in a particular relation to others. "To exist within a social space, to occupy a point or to be an individual within a social space, is to differ, to be different" (Bourdieu, 1998: 9).[3]

Since Bourdieu's problematic regards individuals as being distinguished from one another by their relative position in social space, sociology becomes, in part at least, an exposition of the manner in which the *spacing* between individuals is produced and maintained. This spacing is achieved, Bourdieu claims, through the subject's association, or lack of association, with various forms of cultural and economic "capital." The things we do (our social practices) and the things we acquire or lose (our social capital) determine who we are. "[A]t every moment of each society, one has to deal with a set of social positions which is bound by a relation of homology to a set of activities (the practice of playing golf or piano) or of goods (a second home or an old master painting) that are themselves characterised relationally" (Bourdieu, 1998: 4–5). Individuality is thus a matter of "distinction" (Bourdieu, 1984) according to each subject's values, tastes, knowledge, finances, and so on, a distinction in relation to the patterns of similarity and difference in "social space," a space that is an "*invisible reality* that cannot be shown but which organises agents' practices and representations" (Bourdieu, 1998: 10, my emphasis).[4]

While it will be necessary to spell out in some detail how Bourdieu's social topology might inform an ethics of place, his conception of social space as an "invisible reality" links him directly to the critique of the modernist cult of tangibility. (See chapter 7.) Indeed, Bourdieu demands that we recognize that "the real is relational" (Bourdieu, 1998: 3). We need, he argues, to break with the substantialist mode of thinking that "inclines one to recognise no reality other than those that are available to direct intuition in ordinary experience." We must "apply to the social world the relational mode of thinking that is that of modern mathematics and physics, and which identifies the real not with substances but with relations" (Bourdieu, 1990: 125).[5] Recognizing the relevance of this relationality, whether in the case of subjects or objects, requires us to be sensitive to context, to understand *where* things come from and how they *stand in relation* to each other. In other words theory must make explicit the spatial metonymy that has, for good reason, infused much social theory since its inception. Sociology, whose object of study is, according to its individualistic detractors, so insubstantial, has (consciously or unconsciously) had to emphasize the relational composition of reality. We have already seen (in chapters 2 and 3) that this is so despite Marx's overt materialism and Durkheim's positivism. For, as Bourdieu points out, even "[t]he "social reality" which Durkheim spoke of is an ensemble of invisible relations, those very relations which constitute a space of positions. . . . Sociology in its objectivist moment is a social topology . . . an analysis of relative positions and the objective relations between these positions" (Bourdieu, 1990: 125–26).

Bourdieu and Epistemology: Putting Theory into Place

Bourdieu's work attempts to overcome the structure/agency dichotomy in two ways. First, he attacks the epistemological perspectives he associates with each of these conceptions. He claims that structuralists and phenomenologists conceive of their theoretical activities as giving objective and subjective accounts of society respectively. Insofar as they do this Bourdieu claims that they have fundamentally misunderstood their theoretical relationship to their "objects" of study. Second, he posits a theory of social production and reproduction that neither privileges the individual social actor nor reduces her to merely a support of the social structure. His aim is "to construct the theory of practice, or more precisely, the theory of the mode of generation of practices, which is the precondition for establishing an experimental science of the *dialectic of the internalisation of externality and the externalisation of internality*, or, more simply, of incorporation and objectification" (Bourdieu, 1990: 72). Bourdieu thus seeks to combine a reflexive account of the production of theory with a nondeterministic account of the interpellation of the subject, an account that is, at the same time, a relational critique of substantialism posed in terms of a spatial metaphorics.

Bourdieu's epistemological critique begins with an analysis of those "phenomenological" accounts of social and anthropological research that privilege the voice of the subjects themselves. Here respondents are regarded as being able to give an authoritative account of their own social activities that would otherwise be inaccessible to the researcher. But, Bourdieu claims, to accept these explanations at face value is to believe that the social world is entirely transparent to those occupying it, and to hold that people can give a "true" account of the motivations behind their every action. Neither of these "subjectivist" assumptions can, he claims, be epistemologically justified since the subject is socially constituted within and by a symbolic and cultural order that is far from transparent to those who have no choice but to occupy a place within it.

Second, Bourdieu criticizes those accounts that see anthropology, sociology, and so forth as providing objective and true accounts from the supposedly privileged point of view of a *scientific* observer. Bourdieu associates these "objectivist" pretensions to uncover the underlying structures of social organization with scientific structuralism, for example, the work of Lévi-Strauss, Althusser, and others, and views such structuralism as fundamentally flawed (Levi-Strauss, 1968). While an objectivist epistemological perspective forms an inevitable stage in the break from phenomenology, a break that is necessary before one can start to understand events in their wider social and historical context, Bourdieu believes that structuralism wrongly retains a positivistic and representationalist belief in the objective position of the scientific observer. The greatest mistake of objectivism is to constitute "practical activity as an *object of observation and analysis, a representation*" (Bourdieu, 1990: 2). In other words, this positivistic model of theory formation regards the researcher as someone who *stands apart* from their object of study, someone who can gain an objective overall impression of the true state of things precisely because of their *dis-*

tance from the material they observe. Their observations are not supposed to impinge upon that which is observed but simply to *represent* it in its true form in a neutral theoretical language. Theory is thus envisaged, in terms of Richard Rorty's metaphor, as a "mirror of nature" (Rorty, 1980), a more or less perfect *reflection* of that which it seeks to map.

But Bourdieu claims that theory does not *represent* objects. It is rather an expression of the practical interaction of the theoretician with their surrounding environment, a *product* of a particular form of practical activity, that is, Bourdieu's epistemology might be said to be "productivist." (See chapter 3.) Theory cannot claim to maintain a clinical objectivity because the act of knowledge production entails getting one's conceptual hands dirty. Theory entails a particular kind of framing of the world, a *gestalt* that only arises from a particular mode of practical engagement in that world. One cannot simply conceive of theory as a process of uncovering the truth, an *opus operatum*, but must rather see that science (qua theoretical practice) is itself a particular form of life—a *modus operandi*. As such, science can never attain an objective "view from nowhere," but must acknowledge that the knowledge it produces is indebted to its origins in a particular mode of production. Knowledge is relational. It is, as chapter 4 suggested, a construction that expresses the relationship between the theorist and the material of their study. (Always remembering that this contextual constructivism does not mean that knowledge is simply an artifact; it cannot simply be "made up" since it is always beholden to and must incorporate the object's relations to the theorist.)

An example of the effects of structuralism's allegiance to a representational epistemology can be seen in the way structural anthropology *freezes* the practical activities it studies in a timeless and acontextual discourse. While structuralism claims to reveal the underlying contours of the social world, the cartographic representation it produces ignores the temporal context of actual social actions. A map is inevitably *synchronic*, it fixes things into place, and thereby imparts both a false substantiality to certain objects or practices and excludes a relational understanding of what are actually dynamic and dialectical social processes. "Because science is possible only in a relation to time which is opposed to that of practice it tends to ignore time and, in doing so, reify practices" (Bourdieu, 1990: 9). The structuralist project of uncovering the real objective processes and associations underlying social behavior is bound to fail because, whereas phenomenology "excludes the question of its own possibility" (Bourdieu, 1990: 3), structuralism entails a synchronic reduction that cannot capture the relational dynamics of society.

Bourdieu's epistemological critique of objectivism thus focuses upon its inadequate conception of its own activities. He advocates instead a form of "reflexivity" whereby any theoretical field should come to see itself as a historically and culturally specific practice in a dialectical relationship with its subject matter. Theorists must recognize that they are, to some extent, "prisoners" of their own cultural milieu, both of their society in general and of their own particular theoretical and "cultural" position within that society.[6] But, by reflecting on their relation to surrounding social *mores*, the usually unspoken and taken for granted presuppositions and conventions of the social world, the theorist can attain some critical distance on their own theoretical practices and the form and content of the knowledge they produce.

This kind of critical reflexivity is, as I have argued, vital for undertaking a critique of contemporary environmental ethics and ethical theory in general. Bourdieu's reflexive account of theory formation can help us understand how moral theory comes to take on its restrictive axiological form and how we might reconceptualize morality making use of Bourdieu's own spatial and relational problematic. This problematic is doubly useful since, though Bourdieu says little explicitly about ethics, he develops an account of the relations between subjectivity and social norms that "can be understood as a way of escaping from the choice between a structuralism without subject and the philosophy of the subject" (Bourdieu, 1990: 10).

Reason and the Regulation of the Ethical Subject

So far I have suggested that one of the difficulties for theorizing the production of ethical values lies in avoiding either a structural approach that entirely dissolves moral agency or a methodological individualism that dismisses society as an insubstantial epiphenomenon of individuals' activities. Contra structuralism, we must avoid the tendency to see subjects as nothing more than the vectors or hosts of extraneous social and moral values. Values are not ideological viruses injected into the unsuspecting subject during childhood and then surreptitiously incorporating themselves into the individual's "blueprint" in order to ensure their own reproduction. Butler's and Irigaray's work reminded us that a conscience is not a parasitic disease, a weakness that infects an otherwise healthy and autonomous individual, but an expression of a symbiotic relation to others, a give and take upon which the very possibility of autonomy is premised. Similarly, the constitutive nature of this relationship requires that we recognize that, while we might come to criticize certain values held by others in our society, we cannot extract ourselves entirely from our respective ethical fields. To do so would, ironically, entail a loss of identity since our individuality is not a foundational given but as Bourdieu suggests, a matter of "distinction" in *relation* to patterns of similarity and difference in "social space." Our relationship to others is constitutive of our identity. Butler and Bourdieu each in their own way emphasize that there can be no existential freedom for individuals to pick and choose values without social constraint since freedom and constraint are relational, they are inseparable expressions of the individual's ambiguous emergence within, and response to, society.

This relational problematic, which questions the substantialist emphasis on the individual and/or society as pregiven "subjects" and "objects," has important implications for ethical theory. As previous chapters have emphasized, modern ethical theories have tended to assume a subject/object divide mediated by formal rationality. From this perspective, an open and explicit process of reasoning produces rules or formulae that can then provide the basis for objective ethical relations between formally equivalent moral subjects. The moral axiologist claims to unearth these objective rationales for moral behaviors and values, revealing underlying truths that make ethics" previously hidden, traditional, or mysterious mode of operation "transparent" and thereby render previous accounts of ethics redundant. (An understanding of moral philosophy as an *opus operatum* that is clearly analogous to Bourdieu's characterization of structuralism.)

These rationalizations appear to offer a resolution of sorts to debates about structure and agency by providing principles that can then be used to regulate individual moral behavior. The subject is envisaged as an autonomous agent who is deemed to possess a pregiven rational faculty that enables her to decide for herself about the rationale's legitimacy based upon its internal consistency—its logic—and its representational accuracy—its ability to mirror the external world. Having been offered the possibility of rejecting those rubrics that are "irrational" the subject is convinced, or at least always could be convinced, of the rationale's "objectivity."[7] The same "reason" that delineates the ethical field also ensures each subject's compliance with this new objective moral order.

Despite the elegance of this solution, it is ironic that reason, whose role is to dispel the mystery surrounding ethics' operation, thereby dons mystery's mantle itself. It is reified as an objective and transcendental element strangely present in all subjects.[8] This omniscient and omnipresent reality is given the apparently contradictory roles of ensuring the subject's freedom of thought and, at one and the same time, policing objective agreement. It operates both as a mark and guarantor of each subject's autonomy and, simultaneously, as the ground of their collective (human) identity. Thus a substantialism that begins by insisting on the bounded reality of the autonomous subject comes to rely upon a peculiarly insubstantial and intangible principle that knows no bounds, reason—the "invisible hand" of modernity.

These metaphysical speculations might suggest that, once we scratch the surface of our commonsense perspectives, a relational conception of the ethical subject is certainly no more mysterious than that supposed by substantialist and rationalist accounts. A relational approach does however entail radically different conceptions of the ethical self and of the role of ethical theory. For one thing it doesn't place all the explanatory weight on the operation of a formal rationality. In making rationality the sole medium of ethical critique modernist moral theorists suggest that moral development must necessarily walk hand in hand with the development of the subject's rational capacities. Ethical maturity entails being capable of fully rationalizing and expressing the moral logic of one's position. This is, after all, the impetus behind Kohlberg's production of (and Habermas' support for) a developmental model that categorizes morality according to an ontogenetic series of stages each marked by an increasing ability to utilize formal rationality to support one's moral judgments (Kohlberg, 1981; Habermas, 1990). But as Carol Gilligan points out (see chapter 6) one cannot equate being moral with accepting a particular (masculinist) kind of rationalization for morality (Gilligan, 1982). As we have seen, Gilligan argues that moral development has little to do with the inevitable unfolding of a preprogrammed rational ontogeny present within each individual and following the same essential course for all. (A conception that exemplifies the modern/masculine conception of the subject.) Rather, moral development is intimately connected to the development of a relational self, that is, becoming a "member of a network of relationships on whose continuation [we] all depend" (Gilligan, 1982: 30). Bourdieu's, Butler's, and Irigaray's work supports just such a relational conception of the ethical self.

The modernist emphasis on reason also inevitably focuses attention on the subject's making *conscious* moral decisions and the need for an explicit and *objective* rationale for making such decisions. Here, as elsewhere, there is a tendency to fall back into a discourse of subjectivity and objectivity: to see moral "rules" both as consciously formulated by individuals and transparent to them, and as underlying structures recognized and expressed by the theorist. Bourdieu's reflexive sociology can help formulate an alternative, relational, way of understanding the apparent regularity of social relations without recourse to *reifying* such rules.

Against the Rationalization of Ethics: Social Norms and the *Habitus*

Bourdieu suggests that we understand human activity as being ordered by implicit *strategies* rather than as the result of following explicit rules. (Such strategies are akin to "rules" only in a looser Wittgensteinian sense of operating according to one's intuitive grasp of perceived regularities [Wittgenstein, 1988].) Strategies are not to be envisaged as consciously formulated limits on social action, as rubrics to be applied to situations by rote or requiring explicit rational justification, but as ideologically incorporated, open-ended, and flexible dispositions to act in certain ways. The members of any given society usually have no need of explicit rules or rationalizations. Instead they have an implicit feel for the social field in which they exist and an "unconscious" ability to act within the expected bounds of that field without ever following explicit laws on such matters.

This "feel for the game" is incorporated into the individual through her immersion in society. Behavior is not to be understood as driven by hard-and-fast laws but as the product of dispositions "inculcated in the earliest years of life and constantly reinforced by calls to order from the group, that is to say, from the aggregate of the individuals endowed with the same dispositions, to whom each is linked by his [sic] dispositions and interests" (Bourdieu, 1991: 15). This system of dispositions transmitted from generation to generation is referred to as the *habitus*. This generative *habitus* is, Bourdieu explains, a "series of dispositions *acquired through experience*, thus variable from place to place and time to time" (Bourdieu, 1991: 9). It is a form of practical sense that operates without the necessary mediation of conscious thought and that is radically different from the simple application of a set of acontextual abstract rules. In other words, the *habitus* is a dynamic immanent structure that imperfectly reproduces a given set of social relations in the behavioral strategies of members of future generations. Rather than inducing knee-jerk or mechanical reactions to events, it installs creative dispositions bounded by limits imposed by social conditioning but at the same time mediating a whole variety of reactions to what must always in some respects be the unique circumstances in which individuals find themselves. "Action is not the mere carrying out of a rule, or obedience to a rule. Social agents . . . are not automata regulated like clocks, in accordance with laws which they do not understand" (Bourdieu, 1991: 9).

Bourdieu claims that the role of the *habitus* is exemplified in ancient societies where there are very few explicit rules. Such societies are regulated by the reproduction of the *habitus* within a shared but largely unspoken worldview. Ancient societies can operate in this way because they are more culturally homogeneous. We are dealing with community (*Gemeinschaft*) rather than society or association (*Gesellschaft*). Bourdieu refers to the experience of this unspoken worldview as a *doxa*. Traditional societies have a communal *doxa*. "[I]n the extreme case, that is to say, when there is a quasiperfect correspondence between the objective order and the subjective principles of organization (as in ancient societies) the natural and social world appears as self-evident. This experience we shall call *doxa*, so as to distinguish it from orthodox and heterodox belief implying awareness and recognition of the possibilities of different or antagonistic beliefs" (Bourdieu, 1991: 164).

In traditional societies power distribution and social values are, Bourdieu claims, relatively uncontested; they are untheorized and so largely unquestionable, forming the "second nature" of all those living in that community. As Eagleton puts it, paraphrasing Bourdieu, "[w]hat matters in such societies is what 'goes without saying' which is determined by tradition; and tradition is always 'silent,' not least about itself" (Eagleton, 1991: 157). The *habitus* reproduces a form of life and its associated dispositions and values in such a manner that they remain unquestioned and unquestionable, stable over many generations and relatively unchanging. Nonconformity would be rare in such a society since all are inculcated by the same *habitus* and incorporate the same worldview. Ethical values are relatively stable and shared by all members of society in respect of their given roles in that society. There is little need, or possibility, for ethical and meta-ethical speculation. Indeed, Bourdieu claims, there is little need for theory at all in traditional societies. The transmission of the *habitus* occurs through the experience of practices themselves rather than through the medium of theoretical discourse. Bodily communication performs a much more important function—"bodily *hexis*" is incorporated directly into the individual's dispositions. Our deportment, body language, and forms of life are incorporated and reproduced without being theoretically articulated as we are brought up and interpellated into certain communally recognized "niches." "So long as the work of education is not clearly institutionalized as a specific practice . . . the essential part of the *modus operandi* which defines practical mastery is transmitted in practices, in its practical state, without attaining the level of discourse" (Bourdieu, 1991: 87).

All of this changes radically in modern society; the previously homogeneous community is fragmented by continuous and rapid change and by the proliferation of disparate practices. The increasing complexity of society and the increasing specialization seen within it diminishes the degree to which everyday practical life can be shared by all members of that society. Since the values and dispositions that develop within these relatively autonomous fields of society may be radically different, some method of communicating these differences is necessary. As it becomes impossible to inculcate values through direct experience of social practices, the spoken and written word together with other methods of mass communication come to mediate between increasingly isolated individuals. "Theoretical" discourse and formal education systems come to play an increasingly important role in ensuring the efficient interpellation of the subject within the symbolic order. But as theory becomes the locus of ideological transmission, it has to *codify* practices that were previously experienced directly. Theory codifies practical experience, reducing it to clear, simple, basic formulae that, because of their simplicity and generality, are communicable between members of that society. Such formal codification is one way of ensuring at least a minimal degree of communality.

To codify is to come to regulate social practices by formal rules—to *objectify* and rationalize the previously unspoken *doxa* in a juridical discourse—to impose a symbolic order. "Codification is an operation of symbolic ordering, or of the maintenance of the symbolic order" (Bourdieu, 1991: 80). As more and more of the society's activities become objectified in this way, the *doxa* becomes less influential and theory becomes the major site of communication and conflict, where experiences of practices clash with, or agree with, the expression given to or denied to them in theoretical practice. The implicit *doxa* is replaced by an explicit *orthodoxy* that, because it no longer has the unquestioning consent associated with the *doxa*, can be challenged by equally explicit *heterodoxies*. "It is [only] when the social world loses its character as a natural phenomenon that the question of the natural or conventional character . . . of social facts can be posed" (Bourdieu, 1991: 169).

This critical space between orthodoxy and heterodoxy can, of course, prove invaluable to those who wish to challenge the hegemony of a particular set of values but we should not overestimate its critical potential. If Bourdieu is right then codification is a kind of formal rationality that imposes its own discursive limits on thought and action. Its primary role remains, in a Durkheimian manner, that of maintaining the prevailing symbolic order, of ensuring social conformity, and its dream remains one of compelling all to acknowledge reason's omnipotence. Reason too is largely "silent" about the precise manner in which it encapsulates and reproduces a symbolic order within which subjects are compelled to orient themselves. What is more, while codification can provide an extra dimension within which difference (heterodoxy) can be articulated, it also constantly strives to impose its own kind of synchronic uniformity, to freeze things in place. Should we forget the tenuous connections between "formal rationality" and "practical sense" then we are in danger of replacing the real and fluid experiences of ethical subjects with a reified, one-dimensional, and false representation of their activities and values. This perspective complements Weber's own account of the growth of bureaucracies and of the rationalisation of the life-world and adds a Wittgensteinian subtlety to the Frankfurt School's analysis of the hegemony of formal rationality and codified discourse in modern society. (See chapter 1.)

In summary, the picture Bourdieu paints of modern society is one where a codified and explicit logic dominates social relations. This formal logic finds the vagaries of everyday practical sense anathema, and imposes its own quasijuridical definition of reality. It tries to apply its own criteria to *habitual* behavior by claiming to excavate a logic or grammar that underpins everyday life, a logic that is not really there but is a fiction of its own theoretical/practical relationship to the dispositions it observes. This helps us explain how, like other theoretical practices, moral theory comes to reflect and objectify, in both its *form* and in its *content*, the modern social world. From a perspective informed by Bourdieu's work, the axiological and nomothetic forms taken by current moral theory could be regarded as products of modernity's interaction with the world and not as an expression of the ontological structure of the world. The "rules" and "principles" that moral philosophers claim to unearth are effects of uncritically transferring the mode of juridical social regulation prevalent in modern society onto the moral realm. In other words, theorists steeped in our legal-bureaucratic social ethos, unthinkingly tend to exaggerate the extent to which we consciously follow explicit rules. They take rules and laws as models for the operation of social processes because our society is itself rule governed.

Despite their rhetoric, academic philosophers are scarcely more critical than most about the presuppositions that underlie their practice. The philosopher produces formal theoretical structures for determining ethical values because this fits with their ideologically produced expectations about what ethical theory "must" be like. In other words, formal moral axiologies are a result of unreflexive anthropological practices that take no account of their own social origins. Like structuralist varieties of anthropology, modernist moral philosophy mistakenly assumes their theoretical practice takes the form of an *opus operatum* rather than recognizing that it is a *modus operandi*. Modern ethical theory expresses the perceived patterns of moral dispositions in terms of rules that delimit the scope of who or what can be counted as ethically significant, but this does not mean that such rules actually exist. One must not make the mistake of reifying the results of theoretical practice as an objective account of the world. Our moral identity is best understood in terms of inspiring, developing, and sustaining an ethical *habitus* rather than a compliance with codes, rules, or conscious calculations of benefits and losses.

Ethics and the Ecological *Habitus*

Bourdieu's analysis adds depth to our understanding of modern philosophy's obsession with ethical axiologies and the increasing pressure to rationalize environmental evaluation within technical and formal frameworks amenable to bureaucratic manipulation. If we want to move away from modernism, then we cannot ignore the intellectual and moral call to reflect on how our mode of thinking might embody the past traditions, present circumstances, and future hopes of modernity. Reflexivity demands that we try to situate ourselves and our thinking in respect to others, to recognize how our theoretical expressions are inscribed within, and are products of, the practices of our own social formation.

But reflexivity alone will not assure that we escape all of the ideological effects of the symbolic order. No theoretical perspectives can avoid carrying, somewhere within it, the mark of its specific origins and in order to communicate at all discourses must, consciously or unconsciously, reiterate some elements of the dominant symbolic order. For example, Bourdieu's use of a term like "symbolic capital" inevitably suggests that the social spacing of individuals is competitively motivated in an echo of the dominant capitalist ideology.[9] Even Irigaray, whose ethics explicitly advocates developing a language apart from the masculine symbolic order, where women might speak "among themselves," recognizes the paradox entailed in trying to simultaneously escape from and utilize the hegemony of a given language (Davidson & Smith, 1999). After all this language is part and parcel of who we are, of our subjectivity.

While it may be true that Bourdieu is overly optimistic about theory's ability to gain a critical and reflexive distance on itself, and correspondingly underestimates the possibility for conceptual difference in tribal societies, I believe he is certainly right to emphasize the possibilities theory opens up for heterodoxy. "Postmodern" environmentalists like Jim Cheney see the current intimate association between "theory" and "formal rationality" as an indication that the two are irrevocably aligned, that theory per se is indelibly tarred with modernity's/masculinity's brush, that it always entails abstract processes of rationalization and codification. For this reason they advocate a break with theory's present hegemony and a return to contextual and less abstract modes of thought and action.[10] (See preface.) But theorizing takes many forms and although sympathetic toward the need for contextual discourse I have argued that we cannot reject "theory" *en bloc*—rather, we need to recontextualize theory (Smith, 1993). Even if it were possible to revert to an atheoretical primitivism, a "postmodern" equivalent of a mythic past, as Cheney suggests, we should not do so. To give up the very real reflective freedoms that critical theory can produce in favor of submersion in a communal *doxa*, however environmentally friendly this might (or might not) be, would be to abdicate our responsibilities and compromise our freedom to think for ourselves.[11]

If theory can be made to shed its armor of formalization and regulation we might yet recapture a feeling for its myriad possibilities. This theoretical ecdysis might wake us from the frozen one-dimensionality of current moral philosophy and let language once more weave enchantments and create new insights into the value of a truly magical world. As Irigaray argues, we can use language, including theoretical language *ethically* to create a space for difference, to leave "open the possibility of a different language. Which means that the masculine [and modern] would no longer be 'everything.' That it could no longer by itself, define, circumvent, circumscribe, the properties of any thing and everything" (Irigaray, 1985).

In other words, I want to argue that radical ecology needs to develop both a "practical sense" and a "theoretical" (or reflexive) language that can do justice to the idea of an ethics of place, that is, of creating new relations to environmental others. Both this practical environmental sense—an "ecological *habitus*," and this theoretical environmental ethics, have to be present if a genuinely green alternative to the world's otherwise irremediably bleak prospects is to take root and proliferate. Theory is necessary to *articulate* (voice and connect together) an explicit heterodoxy, a counter to the messages carried by modernist/masculinist ideology. Theory in this sense is not an abstract metadiscourse or a discursive reflection of the real order of things but a constitutive part of producing social and environmental change. As chapters 1 and 4 argued, every theory helps to frame and construct a "moral field" by re-emphasizing and re-inscribing elements of the prevailing social relations, giving voice to some aspects of the social (and natural) environment while repressing others.

But theory, by itself, is not enough. Ethics needs to be *heartfelt*, we need to *care* as well as think about what we do to others. In order to have meaning and sustain a (counter-)culture, both "theoretical" and "habitual" aspects of an ethics of place need to be lived and breathed as an integral part of an ecological "form of life."[12] This informal lived relation to one's fellows (of whatever species) and surroundings is precisely what Arne Naess means by a *modus vivendi*. (See chapter 6.)

Ethical Hermeneutics

Perhaps I can develop the connections between *habitus* and ethical theory further, and simultaneously re-emphasize some of environmentalism's implications for social theory, by a short excursus into hermeneutics.

It may initially seem strange to look to what was originally a philosophy of textual interpretation for enlightenment about our ethical relations to the natural world, but the import of philosophical hermeneutics lies in its concern with understanding, and its limits. It recognizes the ineluctable effects of time and place on language and being, and the necessity of reflecting upon these influences, that is, of reflexivity, if any interpretation is to come close to understanding our own and others' situations.

Those associated with hermeneutics in previous centuries, like Wilhelm Dilthey, saw it as a methodology that would allow its practitioners to attain a universalistic perspective on a text, that is, *the true* interpretation of its author's meaning. But twentieth-century hermeneuticists, like Martin Heidegger, Paul Ricoeur, and Hans-Georg Gadamer, have explicitly distanced themselves from such claims (Grondin, 1994). Gadamer argues that we cannot reach some universal (objective or neutral) perspective and this failure is "not due to a deficiency in reflection but to the essence of the historical being that we are. *To be historically means that knowledge of oneself can never be complete*" (Gadamer, 1998: 302). We are temporally (and spatially) limited beings and this situatedness in time and space is not some accidental state of affairs but the very basis of our being. There is no essential human nature that precedes our coming into being. We see things the way we do and are the way we are because of the times and places we exist in. In other words, our culture's "prejudices," in the very broadest sense of the word, are constitutive of our being. Because of this situatedness we have limited *horizons* that place boundaries on our comprehension of self and others.

Modernist philosophy has, of course, recognized the limitations placed upon us by our immersion in different cultures, but has tended to regard its mission as one of transcending, that is, escaping, from such limitations. It has tried to rise above itself in a search for absolute knowledge, to reach a "view from nowhere" that can give its "rational" voice the authority of objectivity. Gadamer rejects this escapist fantasy. We have no access to any mysterious powers or fundamental capacities that might grant us universal insight into the world. Both the form and content of arguments, however rational or objective they may claim to be, are actually products of particular times and places.

Our horizon includes only that which "can be seen from a particular vantage point" (Gadamer, 1998: 302) and there are always other places beyond the horizon, places that we can, as yet, only imagine. Understanding arises only through the meeting of different "horizons" and in transposing ourselves into others' places. This does not mean *replacing* the other with ourselves, reducing their perspective to ours. It means broadening our horizons through the recognition of the other's ineliminable difference to ourselves. In this way epistemology, like ethics, becomes a self/other dialectic, a process of self-formation attained through recognizing and respecting the differences of others. Gadamer explicitly argues that we cannot begin this process of understanding from the hyperseparate self that characterizes modernism but must recognize that "there is an Other, who is not an object *for* the subject but someone to whom we are bound in the reciprocations of language and life" (Gadamer in Grondin, 1994: x, my emphasis).

There is thus a happy coincidence between Gadamer's hermeneutics and the kind of ethical perspective I have been trying to espouse. But these parallels between understanding and ethics go much further, because for Gadamer, the achievement of understanding cannot be fully rationalized; it is not simply a superficial matter of exchanging (trading) ideas. Understanding is not the sole preserve of reason but requires the recognition of a *practical sense*. One needs to have a *feeling* for the other, how they differ from ourselves, of when we have come closest to recognizing their perspective, and so on. Truth is not something absolute that can be imposed from above by force of argument but requires a sensitivity to circumstance. To use Bourdieu's phrase, understanding is not an *opus operatum*, but a mode of being, a *modus operandi*. This ethical feeling is "more than an emotive condition of knowledge. It is one of the forms of relationship between I and Thou. Certainly there is knowledge involved in this real moral relationship, and so it is that love gives insight" (Gadamer, 1998: 233).

In this way Gadamer avoids an absolute distinction between thought and emotion and between ethics and epistemology. We need a feel for the game of knowing just as much as any other game and this feeling requires a sensitivity to the other's situation, a kind of epistemological tact. "By 'tact' we understand a special sensitivity and sensitiveness to situations and how to behave in them, for which knowledge from general principles does not suffice. Hence an essential part of tact is that it is tacit and unformulable. . . . Thus tact helps one to preserve *distance*. It avoids the *offensive*, the *intrusive*, the *violation of the intimate sphere* of the [other] person" (Gadamer, 1998: 16, my emphasis). The point of such meetings of minds and hearts is not agreement nor even compromise, since this also entails compromising oneself in a moral sense, it is rather to come to a *modus vivendi*, a way of living and a way of understanding the other in all their difference. The "fusion" of horizons does not entail an assimilation of positions, a reduction of all to an economy of the Same, but marks a conscious attempt to bring out tensions and learn from them (Gadamer, 1998: 306). This is why this process is always also one of self-cultivation or self-formation.

It is not accidental that, at the beginning of *Truth and Method*, Gadamer refers to the moral theory of Anthony Ashley Cooper, Third Earl of Shaftesbury, something of a forgotten figure in British philosophy. Shaftesbury and his work has been overshadowed by that of his tutor John Locke, archproponent of that same possessive individualism that necessitated the rebuilding of theoretical ethics on the basis of self-interest, a principle with which it is entirely at odds (Darwell, 1995: 181).[13] Shaftesbury's work is written against the tenor of Locke's and is usually referred to as a moral sense theory (Darwell, 1995) since he argues that one must cultivate a *sensus communis*—a moral feeling for one's community. Shaftesbury is not arguing that humans have a pregiven disposition to act altruistically anymore than he thinks that they are predetermined to be selfish. The *sensus communis* is a capacity to recognize the differences and needs of others that emerges from the process of attaining one's individuality and identity within that community. It is a feeling for what should and should not be done to others attained through a process of self-formation—a kind of *acquired* moral taste. "What Shaftesbury is thinking of is not so much a capacity given to all men [sic], part of natural law, as a social virtue, a virtue of the heart more than the head" (Gadamer, 1998: 24).

The similarity to Bourdieu's *habitus* is obvious. Shaftesbury too suggests that in the process of becoming who we are we acquire a moral feel for the communal game that allows us to respond to our fellows in a fitting way. As Gadamer remarks, "[b]oth taste and judgement evaluate the object in relation to a whole in order to see whether it fits in with everything else—that is, whether it is 'fitting'" (Gadamer, 1998: 38). Like Gadamer, Shaftesbury sees the necessity of both conscious thought (reflection) and the *unformulable* moral feeling that facilitates understanding and comprises what we might call the ethical habitus. He argues that moral conduct requires "a sense of right or wrong" (Shaftesbury in Darwell, 1995: 184), but that this sense is a *cultivated* taste—"one that develops through free public critical discourse about the public good" (Darwell, 1995: 186). To this extent Shaftesbury's ethics might also be described as an *antinomian* ethics (see chapter 5) since "[w]hat makes conduct virtuous and virtue obligating appears to have nothing to do with even a reformed idea of the law. Rather, morality primarily concerns what Shaftesbury calls an agent's *affections*" (Darwell, 1995: 182). Codifying ethics is only necessary where people lack a moral sense, where they have not been able to cultivate a relation to surrounding others or when their moral sense does not agree with that of the law makers. Ethics is, at heart, antinomian and anarchic.

We perhaps now have a clearer idea of the possibilities of understanding ethics in terms of the development of an ethical *habitus* and why we must refuse to substitute rules for relations. As Levinas argues, ethics is a primary passion that precedes ontology, giving birth to self and other and steadfastly refusing to reduce one to the other. The *habitus* originates in and is sustained by the constant dialectic of desire and wonder. It is a practical sense that gives us a feeling for what is ethically fitting, formed through the meeting of our horizons with significant others. This *habitus* is not unaffected by argument (since practice is informed by theory and vice-versa), but, as Butler and Irigaray show, theoretical discourses are only one aspect of the symbolic order within which we (quite literally) *find ourselves*.

Language inevitably orders the world, applying systems of classification to it that both constrain and facilitate our coming to recognize parts of our environment as significant others. Theory articulates these taxonomies but the word cannot replace the world. All language only gets meaning through its participation in a "form of life" and it is this form of life and the individual's relation to it that "dictates" what *feels* right and wrong. Ultimately, philosophy cannot impose limits or regulate who or what should count as a significant other and the attempt to do so usually emerges from a very unethical desire to impose a similar form of life on others, to reduce them to the order of the Same.

The Dialectics and Ethics of Place

As chapter 1 illustrated modernism relates a myth of an expanding circle of moral considerability, a history of moral progress in our relations to nature. Both the empirical evidence of environmental devastation and the increasing objectification and commodification of the life-world belie such a myth. And, I have argued, even those modernist moral paradigms with the best of (deep ecological) intentions have done little more than assimilate nature's complex otherness into different but indelibly anthropocentric economies of the Same. Unfortunately, despite their insights, even those figures like Levinas and Irigaray, who have sought to resist or subvert modernism's rationalization of the ethical sphere, have themselves rarely regarded ethics as anything more than the product of entirely human activity.

Such an oversight is understandable insofar as the geography of modernity is primarily human and humanist, the space we occupy is one increasingly made by and for us according to anthropocentric criteria. Modernism marks the apogee of this anthropic instrumentalization of space as everything and everywhere becomes disenchanted and profane and the world is opened up to exploitation.[14] Modernity feeds upon and empties out the differences between places, converting them into standing reserves of objects that are then disembedded, transported, and transformed.[15] The differences that are constitutive of places are thereby eliminated. Each and every place is remade in modernity's own abstract image, reduced to mere coordinates within homogenous space. The "globalization" of modernity means subsuming all difference, all "places," within an instrumental economy of the Same. As Paul Virilio argues, "reduced to nothing by the various tools of transport and instantaneous communication, the geophysical environment is undergoing an alarming diminishing of its "depth of field" and this is degrading man's [sic] relationship with his environment" (Virilio, 1997: 22).

Modernism thus models itself on the abstraction, order, and functionality of Newton's "absolute space." This universal space is devoid of all lingering connections to particular localities and histories including nature's own irregular intrusions. Le Corbusier expresses this ideal in his "The City of Tomorrow," a nightmare of rationalization made, quite literally, concrete; a megalomaniac product of a "machine-age" where "reason has come into her own in company with science" (Le Corbusier, 1971: 244). Here the modern architect is concerned "with the attainment of perfection and with the modern spirit; and one prime necessity emerges clearly WE MUST BUILD ON A CLEAR SITE. . . . To build on a clear site is to replace the 'accidental' lay-out of the ground . . . by the formal lay-out. Otherwise nothing can save us. And the consequence of geometrical plans is Repetition and Mass-production" (Le Corbusier, 1971: 220).[16]

This all too influential ideology, which seeks to eradicate, rather than conserve, the differences between particular places, spawned innumerable epigones, vacuous urban high-rises that *scrape* at the sky, and suburban "human cells" that sprawl ever outwards over landscapes deprived of all distinguishing features. As we have seen, Marc Augé refers to these as "non-places" (Augé, 1995: 94). In the non-place, as in Le Corbusier's city, "[m]ovement is the law of our existence" (Le Corbusier, 1971: 243). All relations to place, to particular localities and relations, are subsumed under the modernist imperative for constant activity, for movement, transformation, and change.[17] Nothing so epitomizes modernity as the bland uniformity of the motorway service station, the shopping mall, or the airport terminal. And, as earlier chapters have argued, the physical and social geography of these contemporary landscapes also mirrors and evokes the "solitary contractuality" (Augé, 1995: 94) of the supposedly autonomous modern subject and a law-bound "moral field." This modernist spatial problematic is the antithesis of the relational conception of ethics and environment I have attempted to elucidate.

Given the pervasive notion of absolute space within modernity then "place" is inevitably accorded a restricted and restrictive definition. Places are the analysands of space, the bounded locations that occupy its otherwise empty terrain, described by Irigaray as envelopes or vessels that can be filled with changing contents, they are a kind of "dimensional entity." Irigaray further argues that, in Tina Chanter's (1995: 158) words, "[w]oman has been treated as the provider of places for men" indeed that "[a]s for woman, she is place" (Irigaray, 1993: 35).[18] And when place is understood in this (masculine) imagery, woman, like place, becomes reduced to a mere vessel, a containing body for man's activities. Place and woman are instrumentalized, treated as the passive receptacles of the activities of those who move within and between them. Place is that which is inhabited, dwelt within, rather then being constitutive of the act of "dwelling." What is more, if woman "is to be able to contain, to envelop, she must have her own envelope" (Irigaray, 1993: 35) she must have her own duly apportioned place and be forbidden to transgress its boundaries. As Gillian Rose argues, absolute notions of space and place are therefore ways of making tangible, containing, and mastering the (feminine) other; "she is imagined as contained, imagined as having a spatiality of impermeable borders, imagined as having 'the solidity of land'" (Rose, 1996: 70).

Nature, like woman, is also reconfigured as a provider of places for modern man's productive activities. She too becomes instrumentalized, defined as an "environment" in the superficial sense of being that which surrounds and contains, the receptacle of the human(ist) subject. The Earth is merely the solid ground that provides the arena and fuel for modernity's gravity-defying flights of anthropic fancy.

Fortunately, there are many who reject this geographic variety of solid mechanics and argue for a different conception of space and place.[19] Geographers frequently make a distinction between the idea of an abstract, absolute notion of "space" as a constant and pre-existing reality, a container of "discrete and mutually exclusive locations" (Smith & Katz, 1993: 75) and a relational space understood as the matrix and product of social (and environmental) practices.[20] Neil Smith and Cindi Katz point out that absolute space appeals to modernism precisely because it claims to provide a neutral ground that is immune from the "decentring and destabilization of previously fixed realities and assumptions" (Smith & Katz, 1993: 80) that modernism itself has induced. (In this sense absolute space is the geographic equivalent of a neutral rationality, a myth that serves to unite a fragmented modern universe by reducing all difference to a formal equivalence that masks the real operation of power.) This "commonsense" notion of space both epitomizes modernist ideology and justifies the rationalized and privatized enclosure of the self and her surroundings. "This space is quite literally the space of capitalist patriarchy and racist imperialism" (Smith & Katz, 1993: 79).

Smith and Katz' work builds on Henri Lefebvre's classic text *The Production of Space* (Lefebvre, 1994). For Lefebvre too space must be envisaged as relational and social, rather than as absolute, geometric, and empty. Relational space is "constituted neither by a collection of things or an aggregate of (sensory) data, nor by a void packed like a parcel with various contents" (Lefebvre, 1994: 27). A "space is not a thing but rather a set of relationships between things" (Lefebvre, 1994: 83). Lefebvre, like Althusser and Irigaray (see chapter 7), is adamant that the substantialist solid mechanics that regards as real only that which is immediately tangible is entirely mistaken. Indeed all social relations and practices however tangible they may be are real only insofar as they are necessarily and intrinsically spatial. As Edward Soja says, "[s]ocial reality is not just coincidentally spatial, existing 'in' space, it is presuppositionally and ontologically spatial. *There is no unspatialized social reality*. There are no aspatial social processes. Even in the realm of pure abstraction, ideology, and representation, there is a pervasive and pertinent, if often hidden spatial dimension" (Soja, 1996: 46). This is why Lefebvre argues that space "subsumes the things produced, and encompasses their interrelationships in their coexistence and simultaneity—their (relative) order and/or (relative) disorder" (Lefebvre, 1994: 73) "space is social morphology" (Lefebvre, 1994: 94).

For Lefebvre then, "([s]ocial) space is a (social) product" (Lefebvre, 1994: 26) and different kinds of society, different modes of being, produce different kinds of space. "[E]very society—and hence every mode of production with its subvarients . . . produces a space, its own space" (Lefebvre, 1994: 31). And, although Lefebvre remains deeply indebted to Marxism, his concept of space is not economically reductive in a narrow sense (see chapter 3), there are innumerable forms of spatial practice.[21] Even within a capitalist mode of production we are "confronted not by one social space but by many—indeed by an unlimited multiplicity or uncountable set of social spaces" (Lefebvre, 1994: 86). Each space has three aspects. First the *social practice* through which social spaces are produced and reproduced. This practice both depends upon and assures that each individual attains a degree of competence, an understanding of their spatially mediated relations to others. Second, "*[r]epresentations of space* [such as that of absolute space] which are tied to the relations of production and to the "order" which those relations impose" (Lefebvre, 1994: 33). Third, particular *representational spaces*, where these spatial representations or codes are displayed, for example, the architecture of the city.

Though Lefebvre does not refer explicitly to *moral* spaces, ethics is, I would argue, one of the key constituents of social space and it too can be understood in terms of particular practices and representations linked to the relations of production (in the broadest sense) of social spaces. It is in this sense that I have spoken of *moral fields* and tried to draw out the connections between the particular social practices of modernity and of a postmodern radical environmentalism together with their very different representations and representational spaces, for instance, the tree house and the law court. (See chapters 5 and 6.) It is in this sense that we are faced with competing ethical architectures, between a nomothetic morality of abstract space and a contextual and relational ethics of place.[22]

However, some problems remain because Lefebvre's theoretical problematic is still tied to an anthropocentric form of productivism. Despite recognizing that nature is the ultimate source of all spatiality, Lefebvre, like Marx, falls into a partial reading that diminishes nature's role in the dialectic (see chapter 3). The activities that produce *social* space are restricted to the premeditated results of *human labour*." There is a rationality immanent to human *products* that is entirely absent from nature's *works*, from trees, flowers, and so on. "[N]ature does not labour . . . it creates. What it creates . . . simply surges forth, simply appears. Nature knows nothing of these creations" (Lefebvre, 1994: 70). This inevitably leads to the instrumentalization of nature reducing it to that which "provides resources for a creative and productive activity on the part of social humanity; but it supplies only *use value*" (Lefebvre, 1994: 70).

Lefebvre's anthropocentric bias also reproduces the modernist distinction between nature (natural space) and culture (social space). He argues that natural space has been almost entirely replaced by social space. "[N]atural space is disappearing [. . . it] has not vanished purely and simply from the scene" but though "[e]veryone wants to protect and save nature . . . at the same time everything conspires to harm it. The fact is that nature will soon be lost to view [. . . and] lost to *thought*. . . . True nature is resistant, and infinite in its depths, but it has been defeated, and now waits only for its ultimate voidance and destruction" (Lefebvre, 1994: 30–31).

It is precisely this separation between nature and culture and the accompanying instrumentalization of the nonhuman environment that radical ecology has sought to challenge by recognizing nature as an active participant in the production of self, society, and our ethical values. Nature may be masked and find itself constantly subject to transformation and abuse but it has not ceased from being part of the dialectic or from being a constitutive part of our values and our understandings of our place in the world.

Perhaps the most obvious way in which environmentalism has challenged modernism's anthropic homogenization of space has been through a re-emphasis of the import of particular natural places in our lives, the most dramatic example of which is bioregionalism. As the editor of a recent collection of articles on bioregionalism stresses, "[w]e are not mere products of our culture and society. We are also the products of the various places and contexts that we depend on" (McGinnis, 1999: 4).

Since its inception in the 1960s bioregionalism has become an influential strand of radical ecology. Two of its earliest proponents, Peter Berg and Raymond Dasman, define it in terms of "[l]iving in place [which] means following the necessities and pleasures of life as they are uniquely presented by a particular site" (Berg & Dasman, 1977: 399) and reinhabiting that site so as to "establish an ecologically and socially sustainable pattern of existence within it" (Berg & Dasman, 1977: 399). Another prominent bioregionalist, Kirkpatrick Sale, similarly describes how "to become dwellers in the land, to relearn the laws of Gaea, to come to know the earth fully and honestly, the crucial and perhaps only and all-encompassing task is to understand *place*, the immediate specific place where we live. The kinds of soils and rocks under our feet; the source of the waters we drink; the meaning of the different kind of winds; the common insects, birds, mammals, plants and trees; the cycles of the seasons . . . the limits of its resources, the carrying capacity of its lands and waters" (Sale, 1991: 42).

Unfortunately, recent debates about place (bioregions) often seem to reiterate those about values (see chapter 4). In countering modernism's anthropic bias, bioregionalists can easily fall into a reductive view of place analogous to some deep ecologists' (mis)understanding of intrinsic values. Some have tended to ignore the advantages of the relational and constuctivist perspective of space offered by theorists like Lefebvre and retreated into a biogeographical objectivism. This reductive perspective envisages "place" in terms of a geographically defined region of absolute space, a relatively fixed and bounded site that determines and dictates the cultural possibilities of those inhabiting its territory. Thus Sale regards the natural features that define bioregions as "the *givens* of nature. [. . . A bioregion] is any part of the Earth's surface whose rough boundaries are determined by natural characteristics rather than human dictates, distinguishable from other areas by particular attributes of flora, fauna, water, climate, soils and landforms, and by the human settlements and cultures *those attributes have given rise to*" (Sale, 1991: 55, second emphasis mine). Once again, and echoing Cheney (see preface), the only authentic human cultures are those where natural places (bioregions) "speak through us."

In this reductive bioregionalism it is nature not culture that produces a sense of place. The bioregion is "a life-territory, a place defined by its life forms, its topography and its biota, rather than by human dictates; a region *governed* by nature, not legislature" (Sale, 1991: 43, my emphasis). What is more it is only "*within* the boundaries of the region" (Sale, 1991: 46) that we can find solid grounds for lasting values in the face of the transient and artificial anthropocentrism of the modern (socially constructed) world. The solution to modernity's ills is "fully knowing the character of the natural world and being connected to it in a daily and physical way provides that sense of oneness, of *rootedness*" (Sale, 1991: 47).

But the irony is that the ethical relation is thereby reduced to a desire for Sameness and solidity, that is, for unity (oneness) and fixity (rootedness). Indeed, in Sale's vision, morality finds itself reduced to a naturalized version of Durkheimian functionalism (see chapter 2). The "entire moral structure . . . would rest on Gaean principles. Oughts and Shoulds would be based . . . on securing bioregional stasis and environmental equilibrium" (Sale, 1991: 120). "Killing, mugging, rustling, rioting, and the like . . . would bring disfavor not because there are statues and codes against them but because they are seen and felt to be disruptive of the normal social flow of the community, threatening its success and even survival as a self-reliant unit" (Sale, 1991: 120–21). Thus ethics once more finds itself reconfigured, diminished, and confined as limits are placed on our moral responsibility, this time in terms of pregiven bioregional boundaries. "The only way people will . . . behave in a responsible way is if they have been persuaded to see the problem concretely and to understand their own connections to it directly—and this can be done only on a limited [bioregional] scale." Within the bioregion ethics in the Levinasian or Irigarayan sense becomes superfluous; "people will do the environmentally correct thing not because it is thought to be the *moral*, but rather the *practical* thing to do" (Sale, 1991: 53). The idea of an ethics that might seek to transgress these boundaries is simply beyond Sale's imagination. "What, after all is supposed to be my ethical response to Japan's harvesting of the endangered beluga whale, even if I happen to know about it—what's it got to do with morality, anyway?" (Sale, 1991: 52).

This is not to say that all bioregionalism takes this parochial form; it does not.[23] As Deborah Tall points out, we need to understand the ambiguity of our multifarious relations to location and community. Mobility is often an important dimension of our freedom, for example, from tradition, oppressive values, and so on. Rather than espouse the geographically restricted rootedness Sale favors, which seems more suited to trees than humans, Tall suggests we should recognize that a "sense of place may, by now, require a continual act of imagination" (Tall, 1996: 112). This act of imagination must recognize that our "community is widely scattered [. . . and we have] only our own brief intensities of common experience to bind us" (Tall, 1996: 107). This is why the image of a harpooned beluga can still move some of us, it is also why it has *everything* to do with morality.

Reductive bioregionalism remains hampered by both its obsession with boundaries and by its undertheorized and uncritical notion of place.[24] Place isn't reducible to bioregion—*or any other kind of region*. Places are the particular products of unique combinations of social and environmental relations. As Gillian Rose argues, places "differ from one another in that each is a specific set of interrelationships between environmental, economic, social, political and cultural processes" (Rose, 1993: 41). An ethics of place must, like an ethics of sexual difference, "reconsider the whole question of our conception of place, both in order to move on to another age of difference . . . and in order to construct an ethics of the passions" (Irigaray, in Casey, 1997: 321). An ethics of place is not a call for a return to rootedness, it is no more limited by topography than it is by modernity's formal rationality, or than it was by the walls of the ancient Greek *polis*. Everywhere and in all ages those in authority attempt to call ethics to (their) order, to make it serve the needs and promulgate the ideology of the powerful. But ethics always resists, it always subverts, because it is that anarchic excess, that love that refuses to be contained.

After-words

Imagine, for a moment, a form of life so very different from that within which we are currently trapped; imagine a life informed by the values of radical ecology where we have begun "to make whole what has been smashed." The raging storm of "progress" has been calmed; it no longer blows us blindly into that unknown future. We sit at our ease in the green shade quietly thinking of "far other worlds, and other seas," stilled by the solace of our green thoughts. What now becomes of ethics? What is left for thinking in a world where philosophical theory can no longer dictate the limits of value? What shape does an ethical discourse take?

I have suggested that all ethical theories play a role in composing moral fields. They influence the manner in which the ethical aspects of a culture become *articulated*—again in that double sense of "fitting with" other aspects of that form of life and "giving voice" to that form of life. In the modern/masculine order ethics' role was clear, it had to provide a quasijuridical discourse that justified the making of decisions at a distance by those with no actual feeling for what was happening. Theory accomplished this by making a virtue out of a necessity, by claiming that ethics too should be subject to formal rationality and that only reason's objectivity could provide a sound basis for maintaining justice in a fragmented society of competitive individuals. The results of this, in the form of utilitarianism's unfeeling calculus and deontology's bureaucratic wrangling, are all too obvious. At best, in the hands of skilled and *concerned* practitioners, such approaches manage to slow or ameliorate some of modernity's worst excesses. But since, as this book has argued, they carry within themselves an acceptance of the self-same principles that underlie modernity, they can provide no real opposition to the destruction wreaked by chainsaw and bulldozer. All too often, by claiming to make the value of the natural environment tangible, they have actually facilitated rather than opposed the environmental holocaust that modernism unleashes.

Radical ecology seeks to reconstitute our ethical relations to natural others, it wants to produce a *sensus communis* that can be inclusive of humans and nonhumans. This requires a practical "ecological" sense that can only come from an awareness, a feeling, for what is fitting with respect to natural places and our nonhuman fellows, and this feeling can, in turn, only come about through practicing and experiencing the desire and wonder that natural others can produce in our lives. Our ethical feeling cannot be grasped by or formulated within abstract moral theory though it can, often to its detriment, be in-formed (i.e., affected) by it. What is more, if ethics is indeed a *lived* relation, then it can not be strictly an academic pursuit, undertaken in ivory towers. Genuine ethical "expertise" comes only from a life that is lived openly and sensitively, not from an abstract ability to manipulate esoteric language. Indeed ethics itself can only flourish in a culture that celebrates and is open to difference. And this is perhaps where radical ecology can help reconstitute the idea of culture itself so that it no longer has to be regarded as a sphere that excludes nature.

Interestingly, Gadamer begins his discussion of the development of the social sciences (*Geisteswissenschaften*) in *Truth and Method*, by tracing the changing meanings of the term *Bildung* (culture).[25] He notes that, with the dawn of the modern age *Bildung* lost its previously close associations with "natural form," for instance, *Gebirgsbildung*—a mountain formation, and became one of humanism's key words, defined by J. G. Herder as "rising up to humanity through culture" (Herder in Gadamer, 1998: 10). As modernity unfolds, *Bildung* becomes conceived of as an "inner process of formation and cultivation" (Gadamer, 1988, 11) and, at the same time, a movement from the particular to the universal. "It is the universal nature of human *Bildung* to constitute itself as a universal intellectual being. Whoever abandons himself [sic] to his particularity is *ungebildet* ('unformed')" (Gadamer, 1988: 12). In other words, though this is not Gadamer's specific intent, we can see how "culture" comes to take on a specifically modern meaning that emphasizes the uniqueness of humanity, the separateness of the individual, and the primacy of intellectual endeavor.[26]

Culture (*Bildung*) becomes progressively more divorced from nature and our relation to nature becomes both less important and increasingly mediated by reason. Yet, despite this, *Bildung* retains within it the idea of an ethical relation to others. For, despite regarding the ontology of the isolated individual as primary, persons remain *ungebildet* (unformed) so long as they have not recognized their constitutive relations to the wider (cultural) world. This is, for example, exemplified in the popular genre of the *Bildungsroman*, that fictional or factual account of the travails of youth in its search for maturity. Such works emphasize the necessity of experiencing and encountering the alien and unimaginable in order to eventually become "world-wise." For Gadamer, becoming part of a culture, becoming "cultured," always requires an understanding that can only come from being open to difference. Self-formation, becoming an individual, is inextricably caught up with the recognition of and respect for different others. "[T]he general characteristic of Bildung [culture]: keeping oneself open to what is other" (Gadamer, 1988: 17).

Perhaps then, radical environmentalism must remind culture of its own origins, of "nature," that alien "other" it believes it has nothing further to learn from. Perhaps we need to remind modernity that a society that destroys nature's diversity (its profuse difference) will remain forever *ungebildet*. We need an ethical relation to nature that can inform both our understanding and our evaluations. A relation of *desire*, of wanting to own or consume nature, is not enough. In such a world we not only devalue "natural" others and disenchant the world but we ourselves become philistines of the first order. Desire is *self*-defeating since "becoming conscious of oneself in desire is also annihilated by the satisfaction of desire" (Gadamer, 1988: 253). As the last chapter argued, desire that is not tempered by wonder leads only to the eradication of difference and to the impossibility of genuine self-hood, of individuality. There are only two possible outcomes in a world, like modernity, dominated by desire. First, we succeed in satisfying our desires but only at the inevitable cost of annihilating all difference, of becoming reduced to a one-dimensional economy of the Same. Second, since difference might prove much harder to eradicate that we had anticipated, we enter upon an endless attempt to satisfy and assuage desire through a spiraling consumerism. The empirical evidence seems to suggest that these are indeed the ways in which modernity is *playing itself out* of existence.

But if desire is tempered by wonder at nature's enchantments and mysteries, its diversity, and its astounding beauty, then this difference might indeed, in Irigaray's words, "constitute the horizon of worlds more fecund than any known to date" (Irigaray, 1993: 5). This "ethics of place" would require us to come to an understanding of our *situation* with respect to the natural world around us, to come to feel "close" to it but also to know how to keep a respectful "distance" when necessary. This "practical sense" of what is required, this ecological *habitus*, will not in any way satisfy modernity's wants. It is not amenable to bureaucratic (mis)representation and accepts no substitute for genuine engagement. The antinomian ethos of this ethics of place is obviously anathema to those who cannot envisage a way of life that is not regulated by rules, calculated in terms of profits and losses, or dominated by the servile acceptance of codes of conduct. What, they will ask, is the point of an ethics that cannot *tell you* what to do?

One answer might be to say that this book has portrayed a meta-ethical position and that as a second-order theory, a discourse on what ethics is, it is not concerned to provide first order normative answers to specific ethical dilemmas. But this would be entirely disingenuous since, from the very start I have made it plain that I do not believe that one can separate *form* from *content* and have expressly claimed that this ethics is expected to support the political program of a radical environmentalism. An ethics of place, an ecological hermeneutics, is not just a meta-ethics—indeed in some ways it is not a *meta*-ethics at all since the very idea of a meta-ethics presupposes that one can divorce theory from practice and privileges the former over the latter.

Is this ecological hermeneutics then merely a novel form of relativism, a repudiation of even the very possibility of objectively specifying what is good and what evil? In one sense it is indeed a kind of relativism,[27] since it constantly calls us to be aware of the *context* of our evaluations. But it is *not* a relativism where "anything goes" and it certainly does make claims about what might count as good and evil. It has from the very beginning advocated a particular "form of life" that, in complete contradiction to modernity, encourages an ethical, rather than an instrumental, relation to the natural world. Accepting this "ethics of place" entails valuing a kind of self-formation that can only occur through "fusion" with one's surrounding environmental horizons (natural and cultural). This "fusion" must be ethical in that it must not seek to colonize or appropriate nature, to reduce it into an economy of the Same, but to sustain its excess, that "never-surfeited sea" (Shakespeare, 1971: 59) of difference. Radical environmentalism envisages a *post*-modern "mode of production" and an ethical dialectic between desire and wonder that would give due regard to nature's differences and activities. Such environmentalism seeks to sustain a space that is not only a social but also a natural morphology and stop that erasure of nature that allows Lefebvre to claim that it is now only experienced "as regret, as a horizon fast disappearing behind us" (Lefebvre, 1994: 51).

Given my critique of attempts to formulate universal rationales as substitutes for a context-sensitive practical sense, then it follows that terms like "right" and "wrong," "good" and "evil" cannot be specified solely according to reason's explicit codes. However, as is appropriate for an ethics of *place*, a spatial metaphorics, a discourse on places, helps us to get a feeling for the meaning of this ethics. If an ethical relation to others "lets others *be*" then it might be envisaged in spatial terms as *giving others room* to develop and not *shaping* their existence solely for our own instrumental ends.[28] So far as is possible, one should seek not to *limit* others' potential or curtail their activities. Put (much too) bluntly, an ethical relation entails respecting and perhaps even facilitating the ability of others to maintain their differences and create their own space for development. On this reading "evil" might be understood as the intentional, thoughtless, or unfeeling intrusion, constriction, or violation of the space of others' being. Of course this claim cannot, and is not meant to, work as a universal principle to be applied to every circumstance. It does not envisage a relation of *application* at all, but one of *constitution*; my claim is that to understand things in this way might help produce an ethical space and an ecological *habitus*.[29]

It is no criticism of this ethics to point out that every action or inaction *inevitably* changes our relations to others in some way, benefiting some and harming others. I fully accept this. The point is that an ethics of place requires that one cultivate a practical sense of what is significant and fitting and when and where it is so. One can only do this by remaining open and sensitive to environmental change. There are no shortcuts in this process of self-formation and no principles or methods that we can apply to discover what is, or should be, ethically significant.[30] The manifestation of ethics, that is, *the emergence into significance of the other*, occurs in myriad ways, sometimes suddenly, sometimes slowly, from the interplay of individual, culture, and nature. The differences we wonder at in these significant others are incorporated, quite literally, into our being and behavior, our individuality. Yet this process is never entirely within our control since values are neither recognized nor apportioned solely through our human labor. Nature too is active in framing and constituting what becomes significant. The brush of air from a bat's wing or the scent of wild garlic draw our attention, they impinge upon our *consciousness* and on our *conscience*.

Such an ethical understanding of our environment exists even in modernity's wastelands. If it is not to be extinguished then we must recognize the limits modernity places on thinking and actively seek to *make a difference* ourselves. Only when we come to *sense* the presence of otherness in and around us, whether in the sand beneath our feet, the hare's leap, or the swallow's soaring flight, will we start to care. Only through care and consideration will the Earth become a place worth living in, a "garden" for everyone to share.

Notes

Introduction

1. Though there are primitivist critiques that regard all aspects of modernity as indelibly tainted and seek salvation in a vision of a premodern society without any division of labour. See Zerzan (1994).

2. Some of the evidence cited by Cheney to support his claim that modernity emerges around 7000 B.C. actually flatly contradicts his thesis—for example, one cited author declares that the modernist unified concept of the "self" developed between the composition of the Iliad and the Odyssey, that is, some six thousand years later!

3. Cheney (1994: 173) reinforces this view. "We in the postmodernist West are only beginning to see such possibilities in language. Postmodernism makes possible for us the conception of language conveying an understanding of self, world, and community which is consciously tuned to, and shaped by, considerations of the health and well-being of individual, community and land and our ethical responsibilities to each. This postmodernist possibility is an actuality in the world of tribal myth and ritual."

4. Ecology is full of examples of the disastrous introduction of exotic species, for example the rabbit in Australia or Chestnut Blight (*Endothia parasitica*) in the United States. See Krebs (1978).

5. In William Cronon's words (1989: 13), "an ecological history begins by assuming a dynamic and changing relationship between environment and culture, one as apt to produce contradictions as continuities."

6. In North America "the animals involved include three genera of elephants, giant armadillos, pangolins and anteaters, fifteen genera of ungulates (deer and antelopes) and various large rodents and carnivores" (Simmonds, 1994: 4). Although Simmonds recognizes that not all of these extinctions were necessarily a direct result of human hunting the evidence for them playing a major role is convincing. See Simmonds (1991).

7. That Cheney refers to male domination and intratribal violence in contemporary tribal peoples as a "deterioration" exemplifies this Arcadian myth. In

saying this, he implies that these tribes enjoyed a past in which these vices were absent.

8. Interestingly there are other parallels between Rousseau and Cheney. In his "Reveries of a Solitary Walker" Rousseau provides a paradigmatic example for that meditative openness that Cheney believes lets the world speak through us: "The more sensitive the soul of the observer, the greater the ecstasy aroused in him by this [natural] harmony. At such times his senses are possessed by a deep and delightful reverie, and in a state of blissful self-abandonment he loses himself in the immensity of this beautiful order, with which he feels himself at one. All individual objects escape him; he sees and feels nothing but the unity of all things. His ideas have to be restricted and his imagination limited by some particular circumstances for him to observe the separate parts of this universe which he was striving to embrace in its entirety" (Rousseau, 1979: 108). This striking passage with its talk of self-abandonment, sensitivity, and reverie could be read as a romantic version of the deconstruction of the self, the rejection of taxonomic boundaries and an openness to the world that allows it to speak through us.

9. Leopold's land ethic states: "A thing is right when it tends to preserve the integrity, stability, and beauty of the biotic community. It is wrong when it tends otherwise" (Leopold, 1949: 224–25). Cheney sees this ethic as in part formed by a *rootedness* in the Sand Counties of Wisconsin, but there is no doubt that it is meant to apply beyond Wisconsin's boundaries. It is therefore a "colonizing discourse," in the sense Cheney gives to this phrase and as a principle abstracted from the practice in which it originated prescribing *one right way* to relate to the world it must also be considered totalizing.

10. On the question of exclusion and resistance within differing localities, see Sibley (1995) and Pile and Keith (1997).

11. There is surely an inconsistency too in suggesting that modern narratives are both "inbred" and "uprooted" and at the same time arguing that they are foundationalist, that is, seek a solid, extralinguistic basis on which to found meaning.

12. Cheney's claim is problematic since it implies a parochialism whereby contemporary cultures are taken as anachronisms, survivals from the evolutionary past of our own society, rather than as separate peoples with their own cultural history. Assumptions of modernity's evolutionary superiority over the primitive are simply reversed, modernity is re-envisaged as a fall from primitive grace. Moreover, by generalizing about primary peoples and their language Cheney is in danger of burying the vast cultural differences between tribal peoples under the weight of a supposed *essential* similarity, namely their possession of contextual discourse.

13. Jenks also emphasizes the continuity between modernism and postmodernism, albeit often in terms of pastiche, nostalgia, or parody.

14. Though "postmodernism" is itself a label applied to thinkers like Baudrillard, Foucault, Derrida, and others, rather than a position accepted by them.

15. Reflexive in a way that Lyotard claims pre-modern "narrative" cultures certainly are not. "It is hard to imagine such a culture . . . undertaking the analysis or anamnesis of its own legitimacy" (Lyotard, 1991: 23).

16. I am obviously not using the term "modernism" to identify a particular cultural movement in art, literature, architecture, and so on. Henri Lefebvre makes a similar distinction between "modernity" and "modernism," arguing that modernism is a form of ideology qua false consciousness. "Thus modernism consists of phenomena of consciousness, of triumphalist images and projections of the self. It is made up of many illusions, plus a modicum of insight." By contrast modernity is, Lefebvre holds, "a reflective process, a more-or-less advanced attempt at critique and autocritique" (Lefebvre, 1995: 1). The obvious purpose of Lefebvre's distinction is to defend a form of modernity now radically redefined as critical theory. While this is understandable, given Lefebvre's Marxism, I am not concerned with labelling modernism "false". I do, however, want to stress its implications in terms of encouraging environmentally and socially destructive aspects of modern culture. I also prefer to use "modernity" in a more general sense to identify a particular social formation containing discourses that can be more or less critical of or complicit with its dominant ideology "modernism."

17. While Rousseau's Romanticism is a modern discourse in the sense that it both arises in and expresses something of the modern condition, to the extent that Romanticism critiques the myth of progress, the concept of the bounded autonomous self etc. it can be read as a variety of countermodernism. Of course most discourses contain pro- and countermodern elements, another reason for not homogenizing all modern or all postmodern discourses.

18. This is why Arran Gare is wrong to make universal claims of the kind that "[p]ostmodernism is 'ecocentric.' It is associated with respect for non-Western societies and cultures, for the previously suppressed ideas of minorities, for nature worship and Eastern religions and for non-human forms of life" (Gare, 1995: 87). The problematic nature of this statement is compounded when Gare immediately follows this with an analysis of Nietzsche and Heidegger as proto-postmoderns.

19. Though this is not clear from Gare's account, which refers to such closed narratives as the "opposite" of grand narratives. While grand narratives, like those of liberal humanism, may well seek to impose a false unity on disparate figures, the unity enjoyed by narratively enclosed communities is by no means always anarchic or utopian. Such communities usually recognize and enforce rigid internal and external moral boundaries that operate to exclude those deemed "other" from important areas of moral consideration.

20. Cheney's failure to negotiate with the culture in which he finds himself is then doubly ironic, because for Lyotard it is this "[multi]storied (but increasingly *unheimlich*) residence" that is postmodernity's material, formal, efficient, and final cause.

21. Lyotard's ambivalence is (in)famous but his utopianism emerges in his call to "arrive at an idea and practice of justice that is not linked to that of consensus" (Lyotard, 1991: 66). That too part of the aim of this work.

22. Similarly in rural Scotland, where I now live, we inhabit an environment without untouched "wilderness." *All* the geography and landscapes are human influenced, yet the land, its history, and its occupants, human and otherwise still finds voices through us. The intuitive obviousness of this culture/nature dichotomy is being broken down since human influence is in any case now so widespread as to preclude the existence of untouched wilderness. Bill McKibben (1990) suggests that, in one sense at least, the advent of the greenhouse effect and ozone depletion mean *The End of Nature.*

23. One does not need to espouse an isolated conception of an autonomous and bounded subject in order to argue that we are, to a degree, self-constructed, internally motivated, and so on. The world does not *just* speak through us.

24. An "axiology" is a value-theory that attempts to delineate categories of things that can or cannot be regarded as morally considerable. On the origins and uses of the term see Findlay (1970). On its use in environmental ethics, see Smith (1991).

25. On the distinction between positive philosophies, which coordinate thought with contemporary social "reality," and critical philosophy, see Marcuse (1991).

26. Though see Schmidt (1971), and it is certainly true to say that the Marxists of the Frankfurt School have (Habermas excepted) generally been more open to taking "nature" into account. See Vogel (1996).

27. On feminist ethics, see especially chapters 6 and 7 and Gilligan (1983). Hekman (1995) provides a useful overview of Gilligan's work and some later developments in feminist ethics are chronicled in Held (1995). For a concise account of feminist standpoint theory, see Harding (1991). Val Plumwood is only one, but certainly the most influential ecofeminist to engage with social theory. See Plumwood (1993), but see also Salleh (1997) and Merchant (1996). On Irigaray see chapters 7 and 8 of this work and Whitford (1991).

28. Such spatial metaphors are increasingly popular in social theory. See for example Lefebvre (1991), and for a selective but still useful overview Silber (1995). Unsurprisingly spatial metaphors have been extensively developed by geographers. See for example Soja (1990), and Keith and Pile (1993). For an extremely interesting attempt to develop a tradition of spatial thinking in regard to the environment, see Macauley (1992).

29. For more on the notion of a "problematic," see Althusser's essay "Contradiction and Overdetermination" in Althusser (1969).

30. The term *habitus* is used by Pierre Bourdieu to refer to the individuals "unconscious" ability to (literally) incorporate and then re-apply certain open-ended strategies in day-to-day life. It might be likened to gaining a "feel for the game" and is explicitly opposed by Bourdieu to the codifed rubrics that predominate in modern societies. See Bourdieu (1991) and chapter 8 below.

Chapter 1
Against the Rationalization of Environmental Values

1. Wittgenstein's term "form(s) of life" serves here to denote the pattern of interconnecting and overlapping social relations and practices that are the medium within which shared meanings, understandings, and values are produced. For a similar interpretation, see Rudder-Baker (1984).

2. To cite just one influential example, R. M. Hare (1952: 3) believes it possible to define a logical distinction between the generalisable applications of "ought" and "good" in their imperative (non-moral) and value-judgmental (moral) senses. In this way he hopes, via the study of moral language, to be able to do "much to elucidate the problems of ethics itself."

3. My linking "modernity" to capitalism should not be taken as an indication that I believe the two to be identical. I use "modernity" to imply a more inclusive and less economically reductionist approach to the culture of our present and recent past. In particular my analysis will argue that "the idea of modernity is . . . closely associated with that of rationalisation" (Touraine 1995: 10).

4. Many of those economists who are more critical of the assumptions underlying neoclassical approaches to environmental evaluation refer to their work as "ecological economics." Ecological economics entails "a value commitment to work for a sustainable society in an ecological sense" (Söderbaum 1999: 162). While this is welcome, many in this field seem to want to order the economy according to principles derived from, or at least supported by, scientific ecology. See for example Gowdy and Carbonell (1999). However, replacing an economic "reality" with an ecological "reality" still fails to take account of the political and social interests being played out in economics. On this issue, see the excellent article by M'Gonigle (1999).

5. Sagoff (1989) has played a prominent role in exposing abuses of cost benefit analysis. However, his analysis makes a distinction between subjective preferences amenable to economic comparison and objective concerns that "involve matters of knowledge, wisdom, morality, and taste that admit of better or worse, right and wrong, true and false" (Sagoff, 1989: 45). Such a distinction is difficult to uphold. It is more important to recognise that, rather than regarding economic rationality, as Sagoff does, as a "neutral" (though often callously indifferent) instrument, it is actually a value-laden methodology dependent upon a particularly modernist conception of human nature.

6. The values that emerge also differ between "willing to pay" methods depending upon how the questions are asked (Reaves, Kramer, & Holmes, 1999). The economists answer is to such variation is to suggest using those methods with the "most desirable response strategies" (ibid.: 377), that is those that reduce protest bids to a minimum. This assumes that such irritating refusals to comply are merely technical barriers to developing an efficient methodology.

7. To be fair, the Pearce report itself does not envisage the *wholesale* application of market forces across the board. For example, the suggestion that pollution permits should be bought and sold is only to apply to levels below that previously determined as safe. The problem lies in the report's tendency to overextend the application of economics in decision-making processes, particularly by the introduction of cost-benefit analyses. To this end perhaps the most revealing diagram in the report is a figure (5.2) labelled "the costs and benefits of cost-benefit analysis." According to this figure there are *only* benefits and no costs to be derived from such a procedure!

8. Of course not all economists are as myopic as Beckerman. The idea that every person is motivated only by self-interest is the focus of a detailed critique by the economist Amartya Sen (1977). Sen argues that moral commitments frequently entail counter-preferential action by individual agents. He thus creates a distinction between moral values and personal preferences. He claims that economics is wrong to reduce the former to the latter. For a more expansive discussion of the assumptions behind modern economics, see Hollis and Edward (1975). Andrew Brennan (undated) has also criticized the use of economics as an overarching framework supposedly able to represent all values and instead advocates a moral pluralism.

9. This distinction between the instrumental and noninstrumental valuation of nature is beautifully and wittily expressed by Jean-Jacques Rousseau (1979: 109–10): "There is one further thing that helps to deter people of taste from taking an interest in the vegetable kingdom. This is the habit of considering plants only as a source of drugs and medicines. . . . No one imagines that the structure of plants could deserve any attention in its own right. . . . Linger in some meadow studying one by one all the flowers that adorn it, and people will take you for a herbalist and ask you for something to cure the itch in children, scab in men, or glanders in horses. . . . These medicinal associations . . . tarnish the colour of the meadows and the brilliance of the flowers, they drain the woods of all freshness and make the green leaves and shade seem dull and disagreeable. . . . It is no use seeking garlands for shepherdesses among the ingredients of an enema." Rousseau was also aware of the dangers of such instrumental evaluation. "This attitude which always brings everything back to our material interest, causing us to seek in all things either profits or remedies, and which if we were always in good health would leave us indifferent to all the works of nature."

10. Pearce (1995: 58) claims that contingent valuation can capture something of our underlying ethical concerns for nature insofar as it distinguishes between its use value and its non-use values. "WTP for non-use may well capture so-called 'ethical' preferences." But the critical point here is not that "morality can't enter into expressions of preference" but that morals are never simply subjective preferences and for this reason moral (and for that matter some nonmoral) values cannot be captured through monetarization.

11. Alternative economic systems are not necessarily any better than neoclassical approaches in terms of their relation to "letting nature be." Marxist eco-

nomics, for example, also depends upon the transformative activities of labour on nature qua raw material. (See chapter 3 for more on this.)

12. Indeed "preference-utilitarianism" actually provides the theoretical basis of most rational choice and neoclassical theorists. (See Sen, 1977.).

13. Some anthropocentric philosophers have argued that we should respect nature simply on the grounds that we have a duty to respect the well being of future generations of humans by not destroying potential future resources. There are a number of flaws in this approach. First, it is not at all clear that we should have any duties to future people. Second, as Derek Parfit (1984) has pointed out, we do not know what these people will be like or want. Third, even if we assume that future people will be just like ourselves, the potential number of such people is almost infinite. This means that any calculus that took their wishes into account might deem that even very minor actions on our part might reduce the potential well-being of millions of people yet to come and so be immoral. For further debate on our environmental duties to future generations, see De-Shalit (1996).

14. Singer is proposing that autonomous rational argument is the primary cause of the historical extension of moral considerability beyond kin and reciprocal altruism to wider society. But we might note that the development of a language complex enough to produce and express such rational arguments might itself require a fairly stable and complex society, presumably including some moral norms. The historical and causal primacy of rational argument is therefore questionable.

15. In connection with the issue of moral extensionism, it is worth noting the importance that has been attached at different times to phrenology, IQ tests, and other "scientific" methods of discrimination. For a lucid account of such scientific prejudices, see, for example, Gould (1981).

16. All quotations are from chapter 6 of Taylor (1986). Emphasis altered in the first quotation.

17. This failure is common to all such attempts that always reintroduce a formal hierarchy of interests that usually begin with self-replicating molecules and end in humans. See for example my (1992) review of Johnson (1991).

18. The description of separate kinds of value theories economic reductionism/resource utilitarianism; axiological extentionism; and extensionalist holism used here are not, of course, the only possible taxonomy. For example, John Rodman (1983) has identified four forms of "ecological consciousness." The first, which Rodman terms "resource conservation," is motivated by identical anthropocentric considerations to my own category of resource utilitarianism. It argues for the preservation of nature only insofar as it is useful for the long-term survival of the human species, the well-being of human individuals and the continuance of civilization. Rodman traces this position to such influential figures as the American forester Gifford Pinchot, who claimed in true utilitarian fashion that forests should be used for the greatest good of the greatest number (of people).

Rodman's second category is entitled "wilderness preservation" and associates wilderness with the production of intrinsically valuable aesthetic and spiritual experiences. Nature has what might be termed a "therapeutic" or perhaps "romantic" value. Here again, the value of nature appears to reside in its production of particular human experiences and, given the well-known vagaries of human tastes, is similarly open to revision.

Rodman's third position is that of "moral extentionism," which includes those systems of philosophy like Singer's that I refer to above as axiologies. However, Rodman includes in this category only those systems that retain an explicit hierarchy that privileges humanity by allocating intrinsic value to explicitly human characteristics like sentience or intelligence. Despite their reliance on an identical methodology, theories like Taylor's that advocate biospheric egalitarianism on the grounds of more abstract and less explicitly human features are not, according to Rodman, examples of moral extentionism. These theories form part of his fourth category, which he refers to as "ecological sensibility."

19. On the Stoics' view of ethics and holism, see chapter 4, section 5 of Long (1974); on Spinoza, see Deleuze (1988).

20. The ecoholist view reaches its apex in the Gaia hypothesis of Lovelock (1979), which sees the Earth as one giant self–regulating organism. Some of his readers have taken *Gaia* as a semimystical concept entailing specific modes of treating the world. Lovelock seem to give at least tacit approval for the mystical interpretation of his work on the pragmatic grounds that it may influence some people to care for their environment. For example, Christians might be persuaded to see the Virgin Mary as embodied in Gaia and thus come to change environmentally destructive practices. However, Lovelock is concerned to stress the scientific nature of his theory and in his later work denies that his theory has any *necessary* ethical implications. He states that "there is no prescription for living with Gaia only consequences" (Lovelock, 1988).

Interestingly, Lovelock's approach no longer places ethical values on "nature" because of its objective characteristics but explicitly recognizes the import of culturally filtered understandings and appropriations of the natural world. This at least acknowledges that our valuations of nature are intimately related to our cultural heritage. What Lovelock seems to forget is that science too is, to some degree at least, a social product. He escapes the criticisms levelled at Callicott only by driving a wedge between the objective "facts" of his science and the cultural traditions and "values" of different communities, that is, by instituting a fact/value dichotomy that sees no necessary relation between the two fields. This, as chapter 4 shows, is problematic.

21. A turn of phrase that, as chapter 7 will show, is reminiscent of Luce Irigaray's concept of fluid mechanics.

22. Equifinality is defined as an ability to reach "a final state from different initial conditions and in different ways" (Von Bertalanffy, in Mathews, 1993: 96).

23. Nor, since I do not want to enter here on a detailed refutation of Mathews' work, is my critique aimed at the inaccuracy of her ecological claims

about ecosystems. Suffice it to say that while many ecologists have wittingly and unwittingly appropriated this cybernetic metaphorics, few if any would regard there as being grounds for regarding ecosystems as having either a telos or an "interest" in their self-perpetuation in the sense Mathews requires. (See, for example, Horn 1981).

24. Mathews provides no adequate criterion for distinguishing a small "natural" ecosystem like a pond from a cesspool since both are reliant upon external additions to maintain their component populations of organisms.

25. In other words we must recognize that the "privilege" granted to the *solid* and *determinate* by modernity still cannot hide the fact that, by itself, substantialism is inadequate for the task of comprehending the world. When people look back at the twentieth century they may see it as the century of relativity, not just in physics, but in linguistics (Saussure), in literature (think of Musil's *Man Without Qualities*), philosophy (the later Wittgenstein), and so on. In each case, whether we speak of quanta, words, or human individuals, their "internal" state is seen to be dependent upon their relations to those significant others who compose their environs. This form of interdependence and relationality is also, of course, what many refer to when they speak of an "ecological" worldview and in this sense there is something of a convergence of thought at the millennium.

26. Though Mathews does not seem to fully realize the extent to which a metaphysical system needs to be part of a what Wittgenstein refers to as a "form of life" in order to have meaning. That is, a worldview cannot just be accepted as the outcome of a philosophical argument but must be lived.

27. It would however be entirely wrong to take the present argument as putting forward "alien otherness" as itself a criterion for moral considerability.

28. I agree with Tester that philosophical debates about animal rights and environmental ethics exhibit a failure of reflexivity in terms of their apparent inability to recognize or make explicit the particular social and historical assumptions that underlie their claims. However, I do not agree that this means that this entails that our evaluations of animals or the environment are *nothing more than* reflections of society, that they can have no value in themselves. (See chapter 4.)

29. Incidentally Darwin was convinced by a series of experiments he carried out using ingeniously shaped leaves that while worms did not "suffer as much pain when injured, as they seem to express by their contortions," they were not only sentient but capable of an elementary form of abstract thought. Indeed a whole section of this work is devoted to the "intelligence shown by worms in their manner of plugging up their burrows" (Darwin, 1881: 64–98).

30. One might extend Bernard Williams comment that utilitarianism exhibits a "great simple-mindedness . . . having too few thoughts and feelings to match the world as it really is" to reductivist axiologies per se whether utilitarian or deontological (Smart & Williams, 1990: 49).

31. I want to be clear here that I am not simply arguing against moral objectivism. While I am convinced that we do not need, and cannot discover,

"objective" values in nature it remains an empirical truth that people do value natural objects for themselves in a manner precisely analogous to our moral valuation of people. (See below, chapter 6.)

32. See for example Käsler (1988), and the essays collected in Whimster and Lash (1987), though Gunther Roth gives a more guarded appraisal of Weber in his essay in this volume.

33. "The term 'formal rationality of economic action' will be used to designate the extent of quantitative calculation or accounting which is technically possible and which is actually applied" (Weber, 1964: 184–85).

34. Kahlberg claims that *Entzauberung* has a very specific context in Weber's discussions of religion, but almost all other commentators express the wider application of the term outlined here (Kahlberg, 1980).

35. This disenchantment is, I would suggest, the "reason" why it seems so difficult for the modern mind to believe in the intrinsic value of the natural environment. (See chapter 4.)

36. Marx too (1973: 84) made this connection between modern Western society and the emergence of our current conception of the autonomous individual. "Only in the eighteenth century, in 'civil society,' do the various forms of social connectedness confront the individual as a mere means towards his private purposes, as an external necessity. But the epoch which produces this standpoint, that of the isolated individual, is also that of the hitherto most developed social (from this standpoint, general) relations. The human being is in the most literal sense a *zoon politikon*, not merely a gregarious animal, but an animal that can individuate itself only in the midst of society. Production by an isolated individual outside society . . . is as much of an absurdity as is the development of language without individuals living *together* and talking to each other."

37. This ambiguity can be seen in the way Weber argued that the sociologist should herself remain "neutral." While their own (substantive) values were relevant to deciding upon a topic of study their actual work should remain neutral and value-free. This seems problematic given Weber's own emphasis on sociology as an "interpretative" science and he could only maintain this position by trying to maintain a distinction between facts and values, the very distinction that his own analysis of formal rationality suggests is implausible. This ambiguity is clearly brought out in Marcuse's (1968) essay on Weber.

38. For an introductory account of the Frankfurt School, see Bottomore (1984). For more detailed accounts, see Held (1980), Jay (1973), and Wiggerhaus (1986). See also Tar (1977) and Hekman (1983).

39. Though this "subjective reason" should not be confused with Weber's more specific use of "subjective rationality" (see above). Nor is Horkheimer's "objective reason" entirely similar to Weber's substantive rationality since Horkheimer seems committed to claiming that it is possible to produce an objective theoretical totality within which one might mediate between the values and goals of society's different value-spheres. Horkheimer also gives a philosophical

rather than a purely sociological account of this process. According to Horkheimer, following the Enlightenment, "objective rationality" turned inward upon itself in a critique of its own pretensions to objectivity. Reason thus becomes a subjective faculty of the mind, a tool for one's individual purposes.

40. Their bleak prognosis led some critics, like Perry Anderson (1984: 88), to view the entire work of the Frankfurt School as a depressing and unproductive chapter in Marxist history. "For, no matter how otherwise heteroclite, they share one fundamental emblem: a common and latent *pessimism.*"

41. In the case of Marcuse this analysis also owes much to his association with Martin Heidegger, whose explicit antimodernism speaks of nature being turned into a "standing reserve." See Heidegger (1993). See also Zimmerman (1990). For detailed analyses of the relation between Heidegger's thought and environmental issues, see Foltz (1995) and Harr (1993). There is more than a touch of irony in that Heidegger's antimodernism led him to support the Nazis, who, despite their mythological and antimodernist rhetoric, took the process of rationalization to its extreme in the concentration camps where even people were, quite literally and appallingly, regarded as nothing more than a resource.

42. See also the text of a speech given just before his death in 1979 (Marcuse, 1992) and Vogel (1996). The recognition and repudiation of the effects of formal rationality is also found among some conservative commentators. Thus Michael Oakeshott (1962: 1–2) picks out almost identical points in characterizing rationalization. He claims that "the Rationalist never doubts the power of his 'reason' (when properly applied) to determine the worth of a thing, the truth of an opinion, or the propriety of an action. Moreover he is fortified by a belief in a 'reason' common to all mankind, a common power of rational consideration . . . he is also something of an individualist. . . . He has no sense of the culmination of experience, only of the readiness of experience when it has been converted to a formula."

43. Of course, the idea that certain ways of doing philosophy might work according to the requirements of formal rationality is not new. Marcuse dedicated an entire chapter of *One-Dimensional Man* to exposing the therapeutic pretensions of the school of analytic philosophy to which most of those environmental ethicists mentioned earlier belong. He specifically recognizes that, in their present context, such modes of thought often "contribute[s] to enclosing thought in the circle of the mutilated universe of ordinary discourse." See "The Triumph of Positive Thinking: One-Dimensional Philosophy," chapter 7 of Marcuse (1991). Philosophy loses its reflexive and critical edge and instead aligns itself with the current order of things—its only job to make words work more *efficiently* or according to certain *codified* rules so that the social order as a whole is spared contradiction and confusion. It is, of course, these very contradictions that formal rationality strives so hard to overcome in order to keep our current environmentally destructive social order intact.

44. This is a much more specific claim than Jim Cheney's (see preface) and one that makes direct links between modernity as a social formation and the form

and content of modern ethics. Cheney is actually (and ironically) in agreement with a philosophical modernism that believes that rationality becomes autonomous; my point is that it is not, it too is an expression of a time and place.

45. "Material" because axiological ethics takes the reasons supposedly underlying any moral action as the substance of its investigation. "Efficient" because we are presumed to have acted on the basis of these reasons. "Formal" because accounts of moral action have to be given in terms of the rationale behind them. "Final" because we are expected to rationally justify what we hoped to achieve by these actions.

46. This, of course, is precisely the manner in which certain philosophers try to insulate their discipline from all extraneous influences. (See the beginning of chapter 2 and Smith, 1998.)

47. This is not to deny that there are differences in the degree to which this formal rationality is present in the frameworks I have discussed. Economic evaluation has, for many social theorists, been both the paradigmatic case of and the cause for such rationalization. But this tendency toward the rationalisation of environmental values even appears in Callicott's more subtle holism. The self-interested individualism justifying formal rationality plays a vital role in his theory since it is this self-interest that is to be expanded to the wider "self" of nature.

48. From a sociological perspective this formalism has much in common with a positivist methodology. It facilitates the reduction of moral complexities to a small number of laws or procedures that supposedly capture the underlying patterns of ethical evaluation just as scientific laws supposedly capture those of the natural world.

49. I have made no mention here of attempts to revive ethical frameworks from earlier times and places, for example, those with predominantly religious orientation or attempts to apply virtue ethics to environmental issues. See for example Frasz (1993). But here too one would need to ask searching questions about these theories' origins in and indebtedness to (premodern) social structures, whether feudal or ancient Greek. Their relevance for radical environmentalism would depend upon both their ability to provide an analysis of the faults of the modern social formation and their ability to support the kinds of society envisaged by radical environmentalism.

50. But see chapter 6 for remarks on feminist projects like the "ethics of care" that also break with this formalism.

Chapter 2
Closed to Nature

1. Initially this might seem like an admission of guilt, especially since they inform us that they consider their own collection to be "a symptom of a growing willingness on the part of contemporary analytic philosophers to examine the nature and history of their own traditions" (Bell & Cooper, 1990: vi). However, the

degree of historical consciousness exhibited by contributors to the book rarely stretches any further than assessing the commensurability of their present interests with those historical figures who were always already predefined as their progenitors.

For example, the opening paper by Prof. John Skorupski is billed as an "attempt to place the concerns and procedures of analytic philosophy within the broader context of contemporary culture and ideas—to see analytic philosophy as both reflecting and contributing to developments in twentieth century thought as a whole" (Bell & Cooper, 1990: x). But even this promise, which already restricts the range of possible influences to "thoughts," rather than other material constraints, is far from being borne out. We find no mention of hermeneutics, structuralism, feminism, postmodernism, critical theory, Marxism; even existentialism and pragmatism are excluded. It is assumed that thinking in the twentieth century has only taken place in the West and even here we find no mention of philosophy's relation to other disciplines like sociology, psychology, and so on. In Bell and Cooper's index, one finds 32 references to Frege but none to Foucault, 13 to Hume but none to Heidegger, 29 to Russell but none to Ricouer, 6 to Gödel but none to Gadamer. Schlick is mentioned but not Sartre, Austin but not Althusser, Moore but not Marcuse. (Incidentally even Adolf Hitler gets one mention.) This is not an attempt to place analytic philosophy within a historical context but to rewrite the history of philosophy as analysis.

2. The boundaries demarcating philosophy in nonanglophone Europe are not necessarily less rigid. See Bourdieu (1988). Those thinkers referred to as "continental" or postmodern philosophers in Britain and the United States are, as Bourdieu makes plain, actually something of a dissident group of scholars so far as the French academic establishment is concerned.

3. For a sociological analysis of some of analytic philosophy's pretensions to disinterested speculation, see Gellner (1968). Richard Kilminster (1989) claims that "philosophers have become defunctionalised as a group as the result of a number of interwoven social processes, one of which is the rise of the social sciences and sociology in particular." He also points out that philosophers have used a variety of "changing claims to disciplinary autonomy" (Kilminster, 1989).

4. This can be clearly seen in the positivism of Auguste Comte, who was responsible for coining the term "sociology" and attempted from its inception to delimit its epistemic scope. (See Comte, 1974.)

5. This dichotomy has recently been called into question by debates over whether sociology needs a New Environmental Paradigm (Dunlap, 1980; Dunlap & Catton, 1991; Laska, 1993; Newby, 1991) and by innovative approaches like that of Ulrich Beck (1992) and Klaus Eder (1996). (See chapter 4.)

6. I use the term functionalist broadly since Habermas' position is again somewhat ambiguous. As a Marxist who recognises profound divisions within contemporary society he is obviously allied in many ways to the critiques of Parsonian functionalism from conflict theory. Nonetheless, as all commentators recognize, he is heavily indebted to Parsons' work.

7. Durkheim refers to "collective representations" as "states of the *conscience collective* which . . . express the way in which the group conceives of itself in its relations with the objects which affect it" (Durkheim in Lukes, 1988: 6).

8. While the primitive *conscience collective* inculcated a respect for the social order, no matter how circuitously, the cult of the individual constantly seems to undermine the source of its own authority.

9. This comment should *not* be read as in any way supporting such projects as sociobiology, which are, in their own way, every bit as disciplinarily reductive. It does though lend itself to supporting those transgressive projects like Luce Irigaray's, which have been badly misunderstood as biologically essentialist in anglophone circles (Irigaray, 1993a, 1993b). See chapter 7.

10. Philosophy commonly makes an analytic distinction between a descriptive "meta-ethics" and an evaluative "normative" ethics. The former tries to define ethics in general rather than justify particular ethical stances and practices. See also the section on "Normative Ethics and Sociology's Disciplinary Boundaries" below.

11. Cybernetics is a term coined by Weiner (1948).

12. Lewis Mumford was perhaps one of the first theorists to fully realize the extent to which technical apparatuses, from clockwork through steam engines to modern computers could operate as metaphors that fundamentally influence the ways in which we come to think about the world (Mumford, 1947). On the impact of cybernetics on our understanding of what it is to be human, see Tomas (1995).

13. This transformation of ethics into a *cybernetic* normativism obviously has a certain theoretical and pragmatic appeal given current predilections for applying cybernetic metaphors to any and all aspects of the life-world. This is why environmental philosophers, like Freya Mathews, who are still in awe of science, find its closeness to the systems theory that dominates current ecological theory inviting. (See chapter 1.) Klaus Eder (1996) develops a sophisticated cybernetic understanding of the relations between nature and culture based on the neofunctionalist work of Niklas Luhmann (see Smith, 1997).

14. This of course was the point made against Parsonian functionalism by conflict theorists like Dahrendorf.

15. It is of some interest to note that according to Benrubi's *Souvenirs sur Henri Bergson*, Bergson thought Durkheim's conception of morality "accurate with respect to the 'closed morality,' but not with respect to the 'open morality'" (Benrubi in Lukes, 1988: 505n). This distinction is useful insofar as it highlights the conservative repercussions of Durkheim's moral theory.

16. His defense of the philosophical discourse of modernity is the prime example of this faith in "modernity" to evolve progressively (Habermas, 1987).

17. As Vogel admits, in Habermas' "writings of the late 1960's, nature appears only as the object of instrumental action. . . . And in *The Theory of Communicative Action* nature is hardly to be found" (Vogel, 1997: 176).

18. Steven Vogel defends Habermas' relevance to environmental ethics on the grounds that nature partakes of the social due to the effects of human transformative activities. But this remains doubly anthropocentric, giving nature no "voice" of its own, since he claims nature "in itself" cannot be known, and suggests that nature's value becomes defensible in discourse only to the extent that it is humans who "*make* the world that surrounds us a good one" (Vogel, 1997: 189).

19. John Drysek (1990) explicitly suggests broadening the conception of communication so as to include relations between human and nonhuman.

20. Since increasing moral density is both the cause of, and solution to, the division of labor in organic society it constitutes, in its own cybernetic terms, a *positive feedback loop*. The integration and interdependence of society's parts is thus an accelerating trend, but one that can only be maintained by the increasing exploitation of the natural environment. See Smith (1995, 1997b).

21. Habermas' relation to functionalism, in the Parsonian sense, is complex. He criticizes Parsons for failing to distinguish between value-orientations that interpellate individuals into the current social order and those that are utopian and critical in seeking to go beyond this order. In this sense he remains within the tradition of the Frankfurt School. However, his criticisms by no means lead him to abandon functional explanations and he retains, if in a modified form, the general outline of Parsons' cybernetic functionalism. See McCarthy (1984).

22. In this sense radical environmentalism has affinities with Walter Benjamin's critical theory. In Habermas' terms (1987: 15) Benjamin "*extends* [. . . our] future-oriented responsibilities to past epochs." We have a duty to remember past injuries and injustices as a form of "anamnesic redemption."

23. Radical environmentalism critiques those humanist ideologies that classify the world in terms of binary oppositions, male/female, culture/nature, and so on. Ecofeminists in particular have excavated a history of these dichotomies that serve to circumscribe membership of society proper, excluding women and nature (Merchant, 1990; Plumwood, 1993).

24. Jürgen Habermas, who borrows from and engages with theorists from many disciplines, might be thought to avoid the charge of imposing disciplinary boundaries. In a sense though his work is intent upon delimiting the boundaries of a version of sociological critical theory. Indeed most of the tensions within his problematic might be seen as due to his trying to occupy and distinguish a territory somewhere between philosophy and sociology. He wants to retain a universalist conception of the communicative grounds of reason at the same time as admitting the influence of differing social (material) conditions. He thus has to delimit a critical theory that defines itself in opposition to both a relativistic sociology and an *Ursprungphilosophie;* "the illusion of pure theory" (Habermas, 1984).

25. There are exceptions to this rule, for example the focus on "moral panics." But here again ethics is understood entirely in terms of deviance from moral "norms."

26. See for example, Mackie (1978). Mackie refers to normative and meta-ethical positions as "first order" and "second order" questions respectively.

27. Social theory has, since its inception, been riven by the seemingly incompatible requirements that it both describe contemporary society and subject that society to an evaluative critique. It is precisely here, in this sociological tension between description and evaluation, that the question of ethics arises and that morality can announce its presence or absence. For it is not just conservative theoretical perspectives (from my perspective those that support current rationalistic tendencies within modernity) that suffer from this dilemma. The motives for *changing* society, for engaging in social critique, are also intimately connected with taking an ethical stance. Critique is motivated by and depends upon values, yet to change society one needs an understanding of its working—one that would seem to require an objective or impartial description capable of seeing society *in toto*. Durkheim, like most social theorists of merit before or since, struggled with this apparently irreconcilable dilemma.

28. Mauss and Beuchat (1979: 3). Much of what follows is indebted to Fox's excellent introduction to Mauss and Beuchat (1979).

29. This is illustrated in the way that the title "social morphology" was given to the sixth and last section, previously entitled "Diverse Items," of the *Année Sociologique* (Durkheim's journal that attempted to *encompass* the social sciences). It was also in this section that the anthropogeographical work of writers like Ratzel found a somewhat uneasy home.

30. The term "rhizome" is used to describe the anastomosing hyphal strands of fungi like the honey fungus *Armillaria mellea*. It is used metaphorically by Deleuze and Guattari (1987) to describe a model of thinking that has no definite root or origins and no fixed form.

Chapter 3
Social Theory, Nature, and the Production Paradigm

1. This is not to say that discourses, like those of utility or rights, are *necessarily* conservative. In certain contexts they can be used to express a radical critique of the prevailing social order. My point is that they cannot provide a radical critique of modernism per se and the rationalization process since they are complicit in extending this process into the moral sphere.

2. Althusser's actual words are "philosophy is, in the last instance, class struggle in the field of theory."

3. The term "monkey-wrenching" comes from Edward Abbey's 1975 novel *The Monkey Wrench Gang* set around the activities of a group of eco-saboteurs who take direct action against environmentally destructive projects. For an appreciation of Abbey's influence on contemporary environmentalism, see Hepworth and McNamee (1996).

4. Althusser's textual comparison of Marxist and Hegelian concepts in his essay "Contradiction and Overdetermination" (1993) applies this notion of problematic. He rejects any idea of transferring the "essential" meaning of a concept to a different theoretical framework merely by using the same word. In this particular case he denies that Marx's conception of the "dialectic" can be *simply* an inversion of Hegel's. The structural role of Hegel's dialectic is different, having an "*intimate and close relation*" (Althusser, 1993: 104) with his world outlook. The concept "dialectic" cannot be simply inserted unchanged, or even *simply* inverted, into a new theoretical framework, rather it is the concept's relations to other parts of the whole framework that gives it the meaning it has.

5. In this sense the term "problematic" has close affinities with what Imre Lakatos refers to as a research program. (See Lakatos & Musgrave, 1972, and chapter 4 of Newton-Smith, 1981.)

6. See especially chapter 6. This work will draw in particular on Lefebvre's (1994) *The Production of Space*.

7. See also Cohen (1978), who defends Marx against this charge.

8. See for example Laclau and Mouffe (1987). The degree to which society has actually altered is of course a fundamental point of difference between Marxists and postmodern theorists.

9. Althusser refers to these inadequate abstractions as "generality I," the reworking of conceptual material that occurs in the process of theoretical labor as "generality II" and knowledge itself as "generality III."

10. Indeed, Althusser is probably one of the most uniformly unpopular theorists of modern times. He retained his membership of the French Communist Party (PCF) throughout the upheavals of the 1960s in an attempt to exhibit his orthodoxy but his theoretical speculations were still regarded by the PCF with deep suspicion. Althusser's structuralist antihumanism led humanist Marxists like E. P. Thompson (who to be honest had only the slightest grasp of Althusser's theoretical work) to declare that he was defending Stalinism and Maoism (Thompson, 1978). Others still attacked him for developing a "highly sophisticated relativism" (Young, 1978: 129). The murder of his wife Hélène following a mental breakdown did nothing to improve his standing with anyone. See, for example Finn (1996: viii), who "proposes links between . . . Althusser's commitment to scientific Marxism and his dead wife."

11. Thus Althusser claims that a subject's "ideas are his [sic] material actions inserted into material practices governed by material rituals which are themselves defined by the ideological apparatus from which derive the idea of that subject" (Althusser, 1984: 43).

12. Althusser argues that there is a radical epistemological break between Marx's early and later works corresponding exactly to the break between the levels of ideology and of theory (scientific knowledge).

13. Though even the term "objectification" is not without unfortunate economistic connotations since a thinker like Heidegger would regard this in terms of "making 'objects' of things," which is precisely the mode of operation of technological/instrumental rationality. Perhaps "realization" might be a better term to use.

14. Lukács (1978: 22) argues that "[a]nimal consciousness in nature never rises above the better serving of biological existence and reproduction, so that ontologically considered, it is an epiphenomenon of organic being." Human labor on the other hand is "a self-governed act" (Lukács, 1978: 35), an "overcoming of animality by the leap to humanization in labor" (Lukács, 1978: 17). Engels too argued that there was an unbridgeable gulf between human and ape. "The specialization of the hand—this implies the tool, and the tool implies specific human activity, the transforming reaction of man on nature, production" (Engels, 1946: 17).

15. There are many other places where it is obvious that the mature Marx retains the notion of objectification, though labor becomes the sole medium in which objectification takes place. For example "[d]uring the labour process, the workers labour constantly undergoes a transformation, from that of unrest [*Unruhe*] into that of being [*Sein*], from the form of motion [*Bewegung*] into that of objectivity [*Gegenständlichkeit*]" (Marx, 1990: .296).

16. One could, of course, argue that selective breeding is a dialectical process between nature and humanity. But this is to miss the point, which is that Marxists emphasize the causal and premeditated role of *human* labor in this dialectic. Thus Engels (1946: 281) states that "the hand is not only the organ of labour it is also the product of labour" evolving by inheriting the characteristics derived from the labor of previous generation. (This is, incidentally, an idea of evolution that despite Engels' genuflection to Darwin, owes more to Lamarck.)

17. In Engels' words (1946: 291), "the animal merely *uses* external nature and brings about changes in it simply by his presence: man by his changes makes it serve his ends, *masters* it. This is the final, essential distinction between man and other animals, and once again it is labour that brings about this distinction." Though interestingly Engels follows this remark with an "ecological" insight that suggests that this mastery of nature does not always go according to plan. "Let us not however flatter ourselves overmuch on account of our human conquest over nature. For each such conquest takes its revenge on us . . . it has quite different, unforeseen effects which only too often cancel out the first. The people in Mesopotamia, Greece, Asia Minor, and elsewhere destroyed the forests to obtain cultivable land, never dreamed they were laying the basis for the current devastated condition of these countries by removing, along with the forests, the collecting centers and reservoirs of moisture" (Engels, 1946: 292).

18. Although interestingly Marx (1990: 493) does state in a footnote that "Darwin has directed attention to the history of natural technology, that is, the formation of the organs of plants and animals, which serve as the instruments of production for sustaining their life." He is thus willing to apply the metaphor of

production to nature at least when considered separately from human purposes. Engles does occasionally allow nature to play an active role in the dialectic—see note 17 above.

19. Thomas Mun quoted in Marx (1990: 649, fn. 5). The additions in square brackets are Marx's own.

20. To this extent Marx also reiterates many of the historical and cultural prejudices of his day, regarding certain other cultures as "primitive," uncivilized, and lacking the motivation of a work ethic.

21. This echoes a comment made by Seyla Benhabib (1992: 69); both liberalism and Marxism, she says, "share the Promethean conception of humanity in that they view mankind as appropriating itself through the process of changing external reality."

22. It might be possible to hold that the increase in environmental awareness and the consequent re-evaluation of nature is an ideological by-product of capitalism's "distortion" of the dialectic. Here people's concerns for nature are explained away as ideologically induced errors. This would seem to suggest that people's concerns for nature as anything other than a resource would disappear altogether in communism.

23. Grundmann (1991, 110–11) includes a selection of other choice quotations from Marx to dispel any lingering notion that he might have been harboring any repressed concerns for our treatment of nature in itself.

24. There are plenty of neo-Malthusians around but they are much more closely associated with "managerial" approaches to ecological problems and the strengthening of current political structures rather than their overthrow. The classic example is Hardin (1968).

25. Andrew Collier (1994: 9) mentions the facts that the Soviet Union and Eastern Europe were forced into "catching up" with their capitalist neighbors, that as nation-states they competed on the world markets and militarily, and that their version of centralized planning was too remote. He sees no irony in simultaneously claiming in the next paragraph that "some planning needs to be legislated and policed on a worldwide scale."

26. In the same issue Ward Churchill (1992: 213–33) points our that Marx's ideas about "progress" and the labor theory of value are explicitly Eurocentric.

27. For an account of his earlier relationship with socialist politics, see Bahro (1982).

28. He concludes, "a theory committed to the paradigm of production can say nothing" about how the rationality he believes inherent in all communicative activities comes to exist (Habermas, 1987: 82).

29. Baudrillard's point is exemplified in Marx's *Grundrisse*, where he goes so far as to take Adam Smith to task for seeing work as a curse—though of course Smith was referring to forced labor. See McLellan (1971: 145–49).

30. Of course Marxists are aware of the dialectic between use and exchange values but nonetheless seek to retain a notion of human needs that remains outside of the system of exchange, for example, certain basic needs that can be defined universally, for all humans in all times and places. This is of course the fundamental form of Marxist humanism and most Marxists would regard this transcendental ground as providing a necessary vantage point for sustaining a critique of capitalism's distorting influences.

31. The transformative process exists only to gratify human needs as material is worked over again and again "until at last it acquires a form in which it can be the direct object of consumption, in which the consumption of the material and the *abolition of its form* results from its enjoyment by man [sic], and in which its transformation is its utilization" (Marx in Schmidt, 1971).

32. Baudrillard's emphasis.

33. To argue this point fully might take a book in itself since their are innumerable definitional ploys that can and have been exercised to set up an absolute distinction between humans and animals, the use of language, concepts, reflection, even an opposable thumb. One can however turn Marx's argument on its head since in all of these cases we by no means deny human status to those people who lack these abilities. My point is that the existence of a qualitative distinction between human labor and the activities of other animals is dependent upon a residual anthropocentrism of exactly the same kind utilized in the ethical axiologies discussed in chapter 1. This anthropic axiology begins with humanity and then moves "outwards" to see what can or cannot be counted as like us.

34. This does not mean, as is sometimes claimed, that Baudrillard espouses a variety of idealism since he frequently emphasizes the "reality effects" of the process of simulation.

35. *Différance* (with an "a") "is the systematic play of differences, of traces of differences, of the *spacing* . . . by which elements refer to each other" (Derrida in Sturrock, 1979: 165).

Chapter 4
To Speak of Trees

1. "Deep ecology" is a term coined by Arne Naess (1972: 95–100). See also Devall and Sessions (1985) and Tobias (1984).

2. There is now an extensive philosophical literature regarding the meaning of "intrinsic" and "inherent" values (Taylor, 1986: 71–76). O'Neill (1993) provides a comprehensive if somewhat partisan summary of the debate. On moral standing, see Stone (1974), Goodpaster (1978), and Brennan (1984).

3. Although interestingly, having recognized that this resource-centered approach is easily labeled as an exploitative expression of commercial greed, logging corporations now emphasize, through massive public relations exercises, the supposed continuity of company and ecological interests. For exam-

ple, a Western Forest Products' (WFP) guide to logging roads and recreation areas states, "forests are more than a resource—they are our heritage and our future. Responsible forest stewardship is essential to our survival and success as a company. WFP is committed to sustainable development on forest lands in our care for all British Columbians—forever" (Western Forest Products Limited, 1995–96).

4. Gibbons (1994: 83) quotes one logger as saying, "[I]f there is one tree left and it will put food on the table for my wife and kids, I'll cut it down."

5. See for example Moses Martin's statements in the forward to George et al. (1985).

6. The Untouchables are not alone in recognizing nonresource values in trees. As part of many Hindus' funeral rites in Kerala coconut trees are planted above the navel of the deceased and come to incorporate the spirit and take on something of the identity of the dead relative. Here trees and people quite literally merge with each other, although the planted trees have a radically different significance and role from those in wild groves. On the significance of the coconut in other Asian societies, see Giambelli (1998).

7. The classic example is of course Berger and Luckman (1967). This draws heavily on phenomenology as developed by Alfred Schutz.

8. The literature on this topic is truly enormous. For historical examples see Thomas (1984); Schama (1996); Short (1991); Simmonds (1993); in anthropology see Descola and Palson (1996); Milton (1996). Several of the key sociological texts taking a social constructivist approach to nature are discussed below.

9. Cronon (1994). See also Zimmerman (1994), though the contests Zimmerman addresses are those between different sociopolitical accounts of nature such as "social ecology" and "feminism."

10. Interestingly, Soulé and Lease, while claiming their book to be an apolitical interdisciplinary synthesis on the nature of "nature," one that is "neither left nor right" make no attempt to distinguish postmodern deconstructivism from more traditional Marxist materialism: both, they claim, emphasize "economic activities" role in producing ideas of nature.

11. The word directly is important here since ontology is indirectly affected by how we classify things. A fish that is classified as food may find itself rapidly changed from a free-swimming state to being fried.

12. Tester is a case in point since even as an extreme constructivist who claims that the "morality [of animal rights] is rather clever at hiding the utter meaninglessness of animals" (1991: 206), and that "[a]nimals have been made moral subjects . . . *for purely social reasons*" (1991: 207), his entire analysis depends upon the recognition that certain "nice cuddly mammals" (1991: 16), are more easily anthropomorphized than others, as the "animals that are most like us" (1991: 14). In other words despite his sociologically reductionist intentions he still has to depend upon certain accepted claims about the ontology of animals.

13. It is beyond the scope of this book to attempt a full justification of this statement. But Saussure emphasises how signs attain meaning though their systemic relationship to each other rather than through a direct reference to an external objective reality. In this sense all attempts to speak of this "outside" reality are always already caught up in the apparently self-referential and infinite interplay of language. For an interesting discussion of the implications of Saussurian linguistics for speaking of "nature," see Kirby (1997). Deconstructivists' claim that we cannot grasp an unmediated nature might also be seen as an extension of Schutz's phenomenological claim that "[a]ll scientific knowledge of the social world is indirect. It is knowledge of the world of contemporaries and the world of predecessors, never of the world of immediate social reality" (Schutz, 1967). The question of whether such extensions (from knowledge of the "social" to knowledge of the "natural" world, and from social science as an *indirect* method of obtaining *objective* insight into the life-world to all attempts to speak directly of the world at large) are justified is another matter. It should be noted that Schutz himself does not rule out the attainment of *direct subjective* knowledge.

14. There is a certain irony here for while denying the neutral or objective status of the natural sciences many constructivists seem to believe that by not "taking sides" in debates about what the world is "really" like they are themselves remaining neutral and objective! Thus Golinski (1998: 7) claims the constructivist should "maintain a neutral stance toward all the contending claims." This supposed neutrality is belied by the gleeful irreverence they actually show toward science and is difficult to equate with Lynch's statement (above) that they are antifoundationalist. All attempts to describe contending claims will inevitably frame them in a way that is not, and can never be, neutral.

15. In an ironic twist the particular adaptation of Spector and Kitsuse's model that Hannigan (1995: 34) favors has been termed an "ecological model."

16. Burningham and Cooper go on to defend the merits of "strict constructivism."

17. Though Shephard seems much more cautious and less deterministic than Sessions suggests.

18. See also chapter 1.

19. Mathews seems to be espousing a group selectionist analysis that is explicitly rejected by the vast majority of biologists. See Harvey and Greenwood (1978: 142).

20. For example Robert May casts doubt on the links between biodiversity and ecosystem stability often alluded to in deep ecological literature. "[T]his notion has tended to become part of the folk wisdom of ecology. . . . But the empirical evidence is *at best* equivocal" (May, 1981: 219).

21. While science is certainly not solely responsible for our environmental predicament, it is certainly not a disinterested observer either.

22. Through organizations like Earth First! radical ecologists, convinced of nature's intrinsic value, not only *talk* about nature but *act* to halt its destruc-

tion. Perhaps it is this convergence of philosophy and political action that has proven particularly galling to some (but by no means all) Marxists, who, as an item of faith, still believe the sole source of historical change can be found in the apathy of a decomposing working class sold on consumption and to those post-modern academicians who seek to replace politics with facile word play. Both camps ensure that deep ecology has no shortage of detractors each insisting on their own particular way of translating discourses about nature into discourses about culture.

23. See chapter 3.

24. For an analysis of this situation with regard to ethics, see Bauman (1993; 1995).

25. I am obviously not implying that there has to be any formalized equality between all such entities as some deep ecologists claim. (See chapter 1.)

Chapter 5
Environmental Antinomianism

1. Reclaim the Streets http://www.hrc.wmin.ac.uk/campaigns/RTS/

2. But see Ivanson (1988: 143), who regards Foucault as arguing for a "new form of right" and summarily dismisses such quotations as "some rather embarrassing remarks made in the course of a discourse with Moaists."

3. It is often argued that Foucault's work questions the very possibility of attaining freedom in any social formation, no matter how liberating that society may claim to be. In Walzer's (1995: 61) words, "[m]en and women are always social creations, the products of codes and disciplines." This leads many to deem Foucault's work entirely pessimistic. David Sibley (1995: 85) states "Foucault's analysis of social control is depressing. We are left feeling helpless." Yet it seems to me that Foucault, especially in his later works, should be read as showing the need for constant *attention* to our moral situation in order to expand our relative autonomy and to retain and develop the ethical potential we have in all circumstances.

4. Though Coppe denied that he "commonly lay in bed with two women at a time" (Coppe in Hill, 1996: 218). Coppe's pamphlet, "The Fiery Flying Role," led to his imprisonment and was ordered to be burned by an Act of Parliament—the Blasphemy Act of 1650—which sought to ban all works that were "corrupting," "disordering," or "posed a threat of the dissolution of a Humane Society" (Coppe 1987: 6).

5. Thompson (1993: 72) quotes a poem by one Muggletonian

Roar cursed reason roar
You can't disturb me more
For wrongs received
Thy serpent tongue
That with revenge is hung
Tis what thy nature craves.

It is perhaps not entirely fanciful to read the antinomianism of the seventeenth and eighteenth centuries as a precursor to certain aspects of the Frankfurt School critique of Enlightenment rationality. (See Adorno & Horkheimer, 1992.)

6. Glasgow, with many socially deprived areas, has one of the lowest levels of car ownership in Britain. The Pollok Park camp, originally set up in 1993, followed the granting of permission to drive a major road through parkland donated to the people of Glasgow by the founder of the National Trust for Scotland Sir John Stirling Maxwell. The park is also the sight of the internationally famous Burrell Art Museum.

7. See also Brass and Koziell (1997); McKay (1998). The Pollok Park campaign drew on support from a wide variety of local people, environmentalists, left-wing organizations like Militant, and even local schoolchildren. When police tried to make use of a local school playground in an attempt to expel the protesters, schoolchildren walked out of classes and charged the police lines surrounding the camp. As a direct result of this twenty-six locally hired security guards resigned despite the fact that under government regulations they risked losing all state benefits by "voluntarily" making themselves unemployed. On another occasion the Conservative Scottish Minister for Trade and Industry seemingly "threatened protesters with a pick-axe" and subsequently resigned.

8. The Newbury campaign in 1996 was referred to as the "Third Battle of Newbury" in an explicit allusion to the first and second battles, which had occurred during the English Civil War. Here again the campaign drew upon a wide variety of support including that of some of Britain's top mountaineers who joined treetop protests to defy the activities of a commercial climbing team employed by the Sheriff to expel demonstrators. This particular "battle—a clash of philosophies as well as climbing skills" included dangerous treetop escapades and, much to the amusement of the demonstrators, the handcuffing of one of the sheriff's officers to a tree (Vidal, 1996: 3).

9. The Criminal Justice and Public Order Act (1994) was passed by the British Government specifically to extend police powers against nonviolent direct action (NVDA). "Such an increase in the powers available to the police seems to be part of a long-term trend in disciplinary control, which accelerated during the mid-1990's" (Wall, 1999: 126). This trend is continuing despite the change of government with recently proposed legislation threatening to include environmental NVDA against genetically modified crops etc., within the scope of antiterrorist legislation.

10. Foreman's speech given in 1981 is quoted in Lee (1995: 43, my emphasis). Unfortunately, but perhaps not surprisingly, this particular variety of "patriotism" has been interpreted by the FBI as "a harbinger of domestic terrorism" (Foreman in Lee, 1995: 46).

11. "In May 1996, five hundred This Land is Ours activists occupied thirteen acres of derelict land on the banks of the River Thames in Wandsworth, highlighting the appalling misuse of urban land, the lack of provision of afford-

able housing, and the deterioration of the urban environment. The site was destined for the ninth major superstore within a radius of a mile and a half. They cleared the site of rubble and rubbish, built a village entirely from recycled materials and planted gardens. The activists held on to it for five and a half months, until they were evicted by bailiffs acting for the owners Guinness" (Monbiot in McKay, 1999).

12. There is a growing recognition that environmentalists and labor organizations might need to forge alliances rather than being played off against each other by the state and big corporations over the, often spurious, issue of jobs. This is exemplified by the recent tactical alliance in Seattle against free trade and capitalism and in Britain by environmentalists supporting striking Liverpool dockers. This alliance has precipitated extreme reactions by those who perceive their interests to be threatened. Judi Bari, an Earth First! campaigner against logging in the Pacific Northwest trying to build an alliance with logging unions, was severely injured by a bomb planted under her driving seat (Reed, 1999). The FBI accused her of blowing herself up but later dropped the charges.

13. Perhaps ironically she claims his alleged attraction lies in his underlying conformity, the fact that "his mother taught him to be nice. 'If I burp I say sorry'" She feels it necessary to patronize him still further for being "quite staggeringly dim." As John Vidal (1998) points out, Raven's claims about Swampy's ignorance are unfounded. He also points out that Swampy was consistently misrepresented by the media; "[t]he man who, with his peers, was passionately for more open democracy and believed that it should not rest on one vote every five years was accused of being 'anti-democratic' because he said that Westminster politics appalled him."

14. See Anon. (1998), "How to Work on Environmental and Population Issues without Scapegoating Immigrants" http://www.envirolink.org/orgs/ef/scballot.html

15. For example, Haywood (1994: 4) argues that trenchant critics of the scientistic secularism of Enlightenment rationality and their associated institutions like Vandana Shiva are *really* only claiming that the Enlightenment project has not gone far enough.

16. See chapter 2 for an account of Weber's analysis of rationalization and codification in modern societies.

17. This brings us back to that misrepresentation of radical ecology which regards it as a variety of moral reformism, as a movement that seeks to emend (and enlarge) the scope of current legal and moral frameworks to somehow include nature. As chapter 1 illustrated, current ethical theory tries to include nature via a process of axiological extensionism, an "expanding circle" of rationally justified moral considerability that might eventually encompass aspects of the nonhuman world, granting them rights or taking their utility into account. It is these reformists who want to "regulate" our social and natural relations. The argument seems to be that if we find a set of rational criteria that

might justify certain aspects of nature being incorporated into a more comprehensive moral code, if trees have "rights," or if the suffering of earthworms could be quantified in a felicific calculus, then our moral relations with nature could also be regulated and administered by the legal/bureaucratic machinery of the state. For example, Lawrence Johnson (1991) states "[t]hese ideas about . . . moral considerability could serve as the foundation for workable and valuable legal institutions." But such attempts to institutionally "represent" environmental values entirely ignore the antinomian aspect of radical environmentalism. In Foucault's terms Lawrence et al. seek to replace an "instance of collective political elucidation" with a "regulatory instance."

18. "We're just people acting out of conscience, out of our love of this land" (Merrick 1997: 65). On this issue, see Smith (1995).

19. Weber suggests that this legal rationalism, which utilizes a "legally definable relationship of subjection," may have premodern origins in the occidental religious notions of a god who stands above and outside of the world, rather than oriental religions that posited god "within a world which is self-regulated by the causal chains of *karma*" (Weber, 1964: 179). Secular humanism merely replaces the codified commandments of god with the legalistic rules of humanity.

Chapter 6
Against the Enclosure of the Ethical Commons

1. For a similar interpretation of Wittgenstein's "forms of life," see Rudder-Baker (1984) and Davidson and Smith (1999).

2. I use the term "theory" here in its broadest sense to refer to any attempt by a culture to account for the values held.

3. See the preface for my use of the term "postmodern."

4. See also chapter 8 below.

5. Gilligan's is not the only project seeking to radically deconstruct contemporary ethics and reconstitute a breathing space for women out-with its current confines. As the following chapter shows Irigary (1993) regards a feminine ethic as flowing around "desire" rather than "care."

6. Enclosures before the mid-eighteenth century tended to be carried out by private agreement.

7. Tate (1967) states "In the open field village as it existed towards the end of the Middle Ages, in some instances well into the eighteenth century . . . neither county justices nor central government normally had much concern with the everyday conduct of village affairs." See also Yelling (1977).

8. The term might also call to mind Heidegger's notion of "enframing" (*Gestell*). See Haar (1993).

9. "Deferral" might also bring to mind Derrida's use of the term *différance* as insinuating both a "spacing between" and a "postponing until later" (Derrida, 1978: 136).

10. See chapter 1 for an account of the effects of rationalization.

11. The term "interpellation" is Althusser's (1993).

12. Thus, despite its obsession with consequences, utilitarianism *as a system* is not teleological but is an expression of instrumental rationality. It envisages no end or limits to its applicability as a principle. As J. S. Mill remarked, "[t]he corollaries from the principle of utility, like the precepts of every practical art, admit of indefinite improvement" (Mill & Bentham, 1987: 296).

13. Both defenders and critics of Marx alike have (ironically) tended to underestimate the possibilities for new postrevolutionary forms of moral discourse. Debates usually occur within a humanist problematic that focuses on whether or not Marxism actually presupposes implicit notions equivalent to those of individual rights or distributive justice. See Lukes (1985) and the review of this by Soper (1987). See also Neilson (1991).

14. The road movie often makes much of the liberty to be found in constant movement but even here the road eventually runs out. (Think of Thelma and Louise.)

15. The term *habitus* refers to the individual's "unconscious" ability to act within the expected bounds of a socially given field, to bodily incorporate ideologically transmitted open-ended strategies rather than follow explicit rules. See Bourdieu (1991) and chapter 7 below.

16. Perhaps the highlight of this protest was the intervention of one hundred local schoolchildren who charged through security guards lines to successfully halt tree felling. "Children's Crusade Invades M77 Site" (*The Herald* February 15th 1995: 1).

17. One of "Reclaim the Streets" most successful operations closed a three-quarter mile stretch of the six-lane M41 motorway in West London with a giant street party involving up to seven thousand people. "The party . . . featured sound systems, live music and the mock trial of a car." Protesters even managed to break open the road surface with pneumatic drills hidden under the hooped skirts of thirty-feet-high bagpipe-playing carnival figures on stilts. Trees were planted in the resulting holes. *The Guardian*, July 15, 1996, "Trees Planted in Fast Lane at Road Protesters Party." For an Earth First! appraisal of recent antiroad actions see Anon. (1998: 1–4). Roads protests have probably become the focus for environmental protest in the United Kingdom because of the size of the road-building projects in comparison with Britain's limited surface area. In North America protests have tended to target off-road vehicles and road building in wilderness areas rather than in rural or urban settings. A similar argument about ethical fields and natural places could be made for many other examples of modernity's environmental encroachment, such as the damming and canalization of rivers.

18. Though of course there are large numbers of people involved on the fringes of such protests whose "form of life" might not seem radically different to the social norm. The protests are frequently marked by the great variety of individuals from very different backgrounds who give transient support to the protest. These people often become politicized and ethicized by their experiences. This is typified in the *Guardian* March 1, 1996 report of the locals' comments on attempts to evict the Newbury protesters.

19. Thus, for example, contemporary moral philosophy equates "deep ecology" with biospheric egalitarianism. Deep ecologists are then ridiculed and attacked for granting *formal* equality to insects and humans. They make no attempt to actually look and see what meaning these have for environmentalists in practice. This problem is compounded by a number of philosophers, like Lawrence Johnson (1991) and Paul Taylor (1986), who while regarding themselves as radical environmentalists continue to work unquestioningly within the rubrics provided by modernity.

20. Naess says environmentalists should "[f]ormulate strong, clear expressions of values and norms which the opponent cannot neglect" (Naess, 1989: 65).

Chapter 7
Thin Air and Silent Gravity

1. For Weber's analysis of the relation between instrumental rationality and the disenchantment of the world, see chapter 2.

2. In Margaret Whitford's (1991) words, "[t]he imaginary morphology of western rationality is characterized by: the principle of identity (also expressed in terms of quantity or ownership); the principle of noncontradiction (in which ambiguity, ambivalence, or multivalence have been reduced to a minimum); and binarism (e.g. nature/reason, subject/object, matter/energy, inertia/movement)—as though everything had to be either one thing or another. All these principles are based on the possibility of individuating, or distinguishing one thing from another, *upon the belief in the necessity of stable forms*."

3. Marcuse (1991). For a feminist critique of *Homo economicus*, see Ferber and Nelson (1993) and Beasley (1995).

4. Whitford (1991: 92). Irigaray is, of course, playing on the French *homme* (man) here.

5. As earlier chapters have shown, the dominant forms of modern moral theory, whether utilitarian or deontological, serve this current symbolic order. They are rationalizations that seek to codify and quantify the moral field, circumscribing, defining, and thereby objectifying its elements in order to reconfigure it in modernity's own image. These presumptions are nowhere clearer than in utilitarianism, the moral equivalent of the economization and quantification of values. It precisely mirrors the accounting procedures of the economic sphere, engaging in a

"hedonistic calculus" that claims to impartially *sum* up the pros and cons of any given situation on the basis of a formal individual equality. Alternatively, the deontological approaches of those who attempt to allocate moral standing, perhaps even "rights," has more in common with the rationalization of the juridical sphere noted by Weber in that it sees the answer to moral dilemmas as lying in our possession of explicit abstract criteria by which one can make impartial decisions.

6. Thus Irigaray (1985: 106) states: "women diffuse themselves according to modalities scarcely compatible with the framework of the ruling symbolics."

7. This is not to advocate simple binary oppositions but to emphasize the disruptive potential of that which is repressed in the modern/masculine symbolic order, those *missing elements* that are denied a language in which they can be articulated in their own right.

8. This is analogous to Spinoza, for whom, "God is the immanent, not the transitive, cause of all things" (Spinoza, 1982: pt. 1, prop. 18).

9. The very idea of allocating ideology such a prominent role was regarded as problematic by those Marxists who held to a more substantialist form of materialism. This in its turn meant that Althusser had to justify his position by claiming that, despite its disparate and nebulous nature, ideology was nonetheless material, though its materiality was of a different modality. "[A]n ideology always exists in an apparatus, and its practice or practices. This existence is material" (Althusser, 1993: 40). And again; "*ideas are his [sic] material actions inserted into material practices governed by material rituals which are themselves defined by the material ideological apparatus from which derive the idea of that subject*" (Althusser, 1993: 43).

10. Althusser's position here owes much to Lacan, under whom he studied (see Lacan, 1992).

11. Insofar as the notion of the "imaginary" is, in part, derived from Lacan, even the triumph of socialism would not, by itself, necessitate that ideological distortion ceases. For Althusser, one can only escape from ideology's distortions through the acquisition of adequate theoretical (scientific) knowledge. Althusser claims that Marx's later work provides just such a science of societies.

12. The human essence was typically defined according to what were perceived as masculine traits. See Lloyd (1984) and Nye (1988).

13. Thus, for example, David Hoy (1990) defines Derrida's postmodern antihumanism (among other things) in terms of his rejection of any philosophy of universal essences, especially a universal human essence.

14. This theoretical antihumanism is, Althusser claims, derived from the "scientific" writings of the mature Marx, for example, in *Capital* (1990: 92) Marx states: "Individuals are dealt with here only in so far as they are the personifications of economic categories, the bearers [*Träger*] of particular class-relations and interests."

15. For introductory but detailed accounts of the structure/agency debate, see Giddens (1979) and Callinicos (1989).

16. The notion that ethics precedes ontology is, of course, also fundamental to Levinas' work. See Peperzak (1993).

17. This is what Foucault (1980: 141) means when he declares that power takes "multiple forms," it "*is* 'always already there,' . . . one is never 'outside' it."

18. It is, of course, not accidental that Irigaray chooses to unearth the repressed within Descartes' (the first philosopher of modernity) own works. The "passions" of the soul were subsumed within the instrumental rationality of his method.

19. Irigaray is here paraphrasing Descartes, "it appears to me that wonder is the first of all passions; and it has no opposite, because if the object which presents itself has nothing that surprises us, we are in nowise moved regarding it, and we consider it without passion" (Descartes in Irigaray, 1993: 13).

20. "Separation is embedded in an order in which the asymmetry of the interpersonal relation is effaced, where the I and other become interchangeable in commerce, and where the particular man . . . is substituted for the I and the other" (Levinas, 1991: 226).

21. See "The Asymmetry of the Interpersonal" in Levinas (1991: 215–16).

22. This position too has parallels with the work of Emmanuel Levinas. Indeed Tina Chanter declares (1995: 214) that "Irigaray's work as a whole is profoundly influenced by Levinas' conception of ethics." Levinas argues "[a]gainst the thesis that all truths and values can ultimately be reduced to the transcendental activity of an autonomous subject" insisting on "the irreducible moments of heteronomy . . . a pluralism whose basic ground model is the relation of the Same (*le Même*) and the Other (*l'Autre*)" (Peperzak, 1993: 19). Irigaray's final essay in *An Ethics of Sexual Difference* (1993) is a critical reading of Levinas' *Totality and Infinity*. Unfortunately Levinas' phenomenology effectively limits the "other" to substantial human individuals.

23. This will not of course stop the humanist arguing that there are certain constitutive "drives" shared by all individuals. But as Wittgenstein (1988: 32e) points out, one can always produce an essentialist reading if one insists upon doing so. "But if someone wished to say: 'There is something common to all these constructions—namely the disposition of all their common properties'—I should reply: Now you are only playing with words."

24. "The grand reward was the establishment of an intimacy with a large block of the natural world. The whole Sierran landscape and skyscape came brilliantly alive for him, spoke to him in syllables and sentences so that at last the old distinctions between subject and object, the knower and the known, the animate and the inanimate, man [sic] and the exterior world—all these were dissolved into a mutuality, which is to say a real relationship. Moving through the Sierras, Muir felt he was among friends, in company" (Turner, 1985: 191).

25. Which is not to say that one cannot find passages in Muir's writings that do respect nature's difference. However one can never quite shake of the notion that the difference Muir feels between himself and his contemporaries leads him to look for comfort in closeness to nature. See Turner (1985: 75–77).

26. By the term "moment" here I mean to refer to both the momentary and to momentum, to an "instant" and a "moving power" that turns us about.

27. In relation to the discussion of Jim Cheney and Lyotard's work in the preface, it is those totalizing narratives that would reduce all to a logic of the Same that a postmodern environmental ethics must resist.

Chapter 8
A Green Thought in a Green Shade

1. Indeed Bourdieu's translator Richard Nice goes so far as to claim that Bourdieu's work is best understood as being "written against the currents at present dominant in France, 'structuralism' or 'structural-Marxism'" (Nice in Bourdieu, 1991: viii). Yet, despite Nice's claim, there does remain a genetic and theoretical linkage with structuralism in more than one sense. Bourdieu leans heavily upon the insights that structuralist anthropology has provided and his work emerges out of the same philosophical and anthropological problematics, and the same epistemology of science that so influenced Althusser. Bourdieu has also defined his own approach as a variety of "structuralist constructivism" (Bourdieu, 1990: 122).

2. For a useful introductory accounts of Bourdieu, see Robbins (1991) and Jenkins (1992).

3. Bourdieu is not simply claiming that life is motivated by nothing more than a quest to establish one's distinctiveness from others, but that one's identity is fundamentally relational.

4. It would not be stretching things too far to say that distinction might be regarded as the sociological equivalent of Derrida's textual term *différance*. "This idea of difference, or a gap, is at the basis of the very notion of space, that is, a set of distinct and coexisting positions which are exterior to one another and which are defined in relation to one another through their mutual exteriority and their relations of proximity, vicinity, distance, as well as through relations of order, such as above, below and between" (Bourdieu, 1998: 6).

5. Though this should not be taken too literally since it might be thought to imply a new scientism.

6. And this explains why other members of the theoretician's own academic culture find her explanations convincing.

7. The qualification is necessary here since philosophers do not usually require that those who act according to these moral formulae must necessarily have justified them previously from first principles.

8. This is, of course, precisely what Althusser recognised as characterizing humanism. See chapter 7.

9. It almost goes without saying that despite my critique of rationalization I too will have reiterated aspects of modernism both knowingly, for instance, the metaphor of production, and unknowingly.

10. For an interesting and detailed account of debates about the nature and necessity of theory in analytic and postmodern ethics, see Furrow (1995).

11. Much hangs on one's definition of theory here and on the accuracy of both Cheney's and Bourdieu's characterization of tribal peoples. Contra both, I claimed in the preface that *all* cultures employ some kind of abstract "theoretical" discourse, though the form taken by these varies enormously. In other words, in all cultures, people are consciously involved in language production, the world never simply "speaks through us" (Cheney) in an unmediated fashion, nor are tribal peoples completely without the possibility of challenging authority, that is, of heterodoxy (Bourdieu). The production of explicit taxonomies and conceptual frameworks seems to me to be an integral part of even the simplest social practices and, for that matter, bodily *hexis* remains vitally important in experiencing and communicating values in modern society. See Smith (1995).

12. The term "form of life" is Wittgenstein's and is a way of designating the embeddedness of specific language-games in their wider natural and cultural environment. For a discussion of this in relation to Irigaray's ethics, see Davidson and Smith (1999). This Wittgensteinian perspective would also suggest that returning to the myths of other cultures as ethical archetypes will not work for us.

13. Stephen Darwell (1995: 181) insightfully suggests that another reason why Shaftesbury "has been insufficiently appreciated by anglophone historians of philosophy, especially in this century, is that he appears to care little for the analytical rigor, systematic clarity, and consonance with developing scientific understanding to which English-based philosophy has predominantly aspired in the modern period."

14. "Not only directions in space but the values of the Western world lost their former inviolability. . . . The most material consequence of the loss was a blurring of the distinction between the sacred space of the temple and the profane space outside" (Kern,1983: 179).

15. Anthony Giddens (1996: 22) refers to the "dissembedding mechanisms intrinsically involved in the development of modern institutions." See also Lash and Urry (1994).

16. Even when Corbusier urges that "[w]e must plant trees" (1971: 80), these are to stand in open spaces, without untidy undergrowth as an orderly adjunct to buildings or in rows on streets; they are uniform, characterless ornaments provided only so long as they serve a function in diverting those who find modernism's bleak vision disconcerting.

17. The futurists, modernism's storm-troopers, instituted what almost counts as a new religion of movement and acceleration. In Marinetti's words "[w]e cooperate with mechanics in destroying the old poetry of distance and wild solitudes, the exquisite nostalgia of parting, for which we substitute the tragic lyricism of ubiquity and omnipresent speed" (Marinetti in Kern, 1983: 119).

18. Irigaray's discussion takes the form of a critique of Aristotle's account of place in part IV of his *Physics*.

19. "Common to all of these rediscoverers of the importance of place is a conviction that place itself is no fixed thing: it has no steadfast essence. . . . [N]one of the authors . . . is tempted to undertake anything like a definitive, much less an eidetic, search for the formal structure of place. Instead, each tries to find place *at work*, part of something ongoing and dynamic, ingredient in something else: in the course of history (Braudel, Foucault), in the natural world (Berry, Snyder), in the political realm (Nancy, Lefebvre), in gender relations (Irigaray), in the productions of poetic imagination (Bachelard, Otto), in geographic experience and reality (Foucault, Tuan, Soja, Relph, Entrekin)" and so on (Casey, 1997: 286).

20. Given a subject matter that explicitly includes elements of both nature and culture geography would apparently occupy the perfect academic position for responding to environmental issues. However, the discipline remains divided by intellectual infighting between largely positivist and scientistic *physical geographers* and *human geographers* who must constantly define themselves against what they regard as this unjustifiable "naturalism." In this way the human geographer finds herself in pretty much the same position as the sociologist (see chapter 3) and calls upon pretty much the same intellectual resources and traditions. The proviso is of course that the human geographer more readily employs a spatial metaphorics and pays more attention to the spatial mediation of social arrangements. Many geographers, like sociologists and philosophers, have become increasingly aware of the need to find new ways of mediating between nature and culture without reducing one to the other. See for example Wolch and Emel (1998).

21. Derek Gregory similarly argues that "[a]lthough Lefebvre is concerned to chart the historical succession and superimposition of modes of production of space, his project is not a simple extension of Marx's critique of political economy" (Gregory in Benko & Strohmayer, 1997: 206).

22. Geography and architecture alike have recently become far more aware of this ethical aspect to their respective disciplines. This is exemplified by the arrival of new journals like *Ethics, Place and Environment* and *Philosophy and Geography;* monographs like Harvey (1996) and Harris (1997), and collections such as Proctor and Smith (1999) and Campbell and Shapiro (1999).

23. See for example Thomashow (1999). Although Thomashow still operates with an absolute rather than a relational view of place.

24. Andrew Brennan (1999) contrasts Sale's "homely bioregionalism" with the much more varied and less reductive notions of place proposed by Casey (1993) and Malpas (1994).

25. Gadamer points out that the term *Geisteswissenschaften* is not directly equivalent to the Anglo-American social sciences but was originally a term that arose from the need to translate John Stuart Mill's "moral sciences" into German.

26. In other words *Bildung*, like reason, becomes synonymous with an anthropic definition of humanity. Just as humans are defined as rational animals so it is also our calling to *cultivate* ourselves. Both reason and culture are Promethean gifts that allow humanity to transcend the given (our instinctive, emotional, natural beings) and in doing so both comprehend the world from a universal perspective and guarantee our independence and autonomy through practicing a self-conscious reflexivity. Both too are productivist in the sense that they entail a working on self/world in order to remake that self/world.

27. In exactly the same way that the later Wittgenstein's philosophy of language is also, in one sense, an extreme form of relativism.

28. It is the very opposite of a managerial approach that tries to make nature fit predetermined and usually anthropocentric patterns.

29. Althusser points out that the very idea of application is tied to an instrumental approach. "We all have in our heads the common and convenient notion (in reality an ideological notion) of application as the effect of an impression: one 'applies' a signature under a text, a design on fabric, a stamp on an envelope. An appliqué is a thing that can be posed *on* or *against* something else. The original image of this notion is that of super-imposition-impression. It implies the duality of objects: what is applied is different from that to which it is applied; and the *exteriority*, the *instrumentality* of the first is relative to the second. The common notion of application thus takes us back to the world of *technology*." Constitution on the other hand would suggest that the ethics here expounded "is neither a tool nor an instrument, nor a method . . . but an *active participant*" (Althusser, 1990: 87).

30. It is, however, clear that one cannot rule certain species or beings in or out of ethical bounds because of their similarities or differences to us, as axiologies try to do.

Bibliography

Abbey, Edward (1979). *The Monkey Wrench Gang*. New York: Avon.

Adam, Henry (1976). *Tahiti*. New York: Scholars' Facsimiles and Reprints.

Adorno, Theodor W. & Max Horkheimer (1992). *Dialectic of Enlightenment*. London: Verso.

Advocates for the Environment (1997). "Clearcut Future? The State of B.C.'s Forest, 1997." *Sierra Legal Defence Fund Newsletter* 7: 4–5.

Alexander, Jeffrey C. (1987). "The Dialectic of Individuation and Domination: Weber's Rationalisation Theory and Beyond." In Sam Whimster & Scott Lash (eds.), *Max Weber, Rationality and Modernity*. London: Allen & Unwin.

Althusser, Louis (1969). *For Marx*. Harmondsworth, U.K.: Penguin.

——— (1990). *Philosophy and the Spontaneous Philosophy of the Scientists*. London: Verso.

——— (1993). *Essays on Ideology*. London: Verso.

Althusser, Louis & Etienne Balibar (1986). *Reading Capital*, trans. Ben Brewster. London: Verso.

Anderson, Perry (1984). *Considerations on Western Marxism*. London: Verso.

Anon. (1998). "Direct Action Six Years Down the Road." In *Do or Die: Voices from Earth First!* Brighton: Do or Die Collective, pp. 1–4.

Attfield, Robin (1991). *The Ethics of Environmental Concern*. Athens: University of Georgia Press.

Augé, Marc (1995). *Non-Places: Introduction to an Anthropology of Supermodernity*. London: Verso.

Aylmer, G. E., ed. (1975). *The Levellers in the English Revolution*. London: Thames and Hudson.

Bahro, Rudolf (1994). *Avoiding Social and Ecological Disaster: The Politics of World Transformation*. Bath, U.K.: Gateway Books.

——— (1982). *Socialism and Survival*. London: Heretic Books.

Bakunin, Michael (1973). *Selected Writings*, ed. Arthur Lehning. London: Jonathan Cape.

Baudrillard, Jean (1975). *The Mirror of Production*. St. Louis: Telos Press.

——— (1981). *For a Critique of the Political Economy of the Sign*. St. Louis: Telos Press.

——— (1992). *Selected Writings*. Oxford: Polity Press.

——— (1993). *Symbolic Exchange and Death*. London: Sage.

——— (1994). *America*. London: Verso.

Bauman, Zygmunt (1989). *Modernity and the Holocaust*. Cambridge: Polity Press.

——— (1993). *Postmodern Ethics*. Oxford: Blackwell.

——— (1995). *Life in Fragments: Essays in Postmodern Morality*. Oxford: Blackwell.

Beasley, C. (1995). *Sexual Economyths: Conceiving a Feminist Economics*. New York: St. Martin's Press.

Beck, Ulrich (1992). *Risk Society: Towards a New Modernity*. London: Sage.

Beckerman, Wilfred (1990). *Pricing for Pollution*. London: Institute for Economic Affairs.

Bell, David & Neil Cooper, eds. (1990). *The Analytic Tradition: Meaning, Thought and Knowledge*. Oxford: Blackwell.

Benhabib, Seyla (1992). *Situating the Self: Gender, Community and Postmodernism in Contemporary Ethics*. Cambridge: Polity Press.

Benjamin, W. (1992). *Illuminations*. London: Jonathan Cape.

Bentham, Jeremy (1907). *An Introduction to the Principles of Morals and Legislation*. Oxford: Oxford University Press.

——— (1995). *The Panopticon Writings*. London: Verso.

Benton, Ted (1989). "Marxism and Natural Limits: An Ecological Critique and a Reconstruction." *New Left Review* 178: 51–86.

Berg, Peter and Raymond Dasman (1977). "Reinhabiting California." *The Ecologist* 7.10: 399–401.

Berger, Peter & Thomas Luckman (1967). *The Social Construction of Reality: A Treatise in the Sociology of Knowledge*. Harmondsworth, U.K.: Penguin.

Berman, Marshall (1991). *All That Is Solid Melts into Air: The Experience of Modernity*. London: Verso.

Bernstein, J. M. (1995). *Recovering Ethical Life: Jürgen Habermas and the Future of Critical Theory*. London: Routledge.

Best, J. (1989). "Afterword: Extending the Constructivist Perspective." In J. Best (ed.), *Images and Issues: Typifying Social Problems*. New York: Aldine de Gruyter.

Best, Steven & Douglas Kellner (1991). *Postmodern Theory: Critical Interrogations*. London: Macmillan.

Best, Steven (1995). "The Commodification of Reality and the Reality of Commodification: Baudrillard, Debord, and Postmodern Theory." In Douglas Kellner (ed.), *Baudrillard: A Critical Reader*. Oxford: Blackwell.

Blake, William (1974). *Blake: Complete Writings*. Oxford: Oxford University Press.

Bloch, Ernst (1995). *The Principle of Hope*. Volumes 1–3. Cambridge, Mass.: MIT Press.

Bottomore, Tom (1984). *The Frankfurt School*. Chichester and London: Ellis & Horwood.

Bourdieu, Pierre (1984). *Distinction: A Social Critique of the Judgement of Taste*. London: Routledge.

——— (1988). *Homo Academicus*. Oxford: Blackwell.

——— (1990). *In Other Words: Essays Towards a Reflexive Sociology*. Cambridge: Polity Press.

——— (1991). *Outline of a Theory of Practice*. Cambridge: Cambridge University Press.

——— (1998). *Practical Reason: On the Theory of Action*. Cambridge: Polity Press.

Bradford, George (1989). *How Deep Is Deep Ecology*. Ojai, Calif.: Times Change Press.

Brass, Elaine & Sophie Poklewski Koziell (1997). *Gathering Force: DIY Culture—Radical Action for Those Tired of Waiting*. London: The Big Issue.

Brecht, Bertolt (1959). *The Selected Poems of Bertolt Brecht*. New York: Grove Press.

Brennan, Andrew (undated). "Environmental Ethics and Moral Rationality." Draft paper.

——— (1984). "The Moral Standing of Natural Objects." *Environmental Ethics* 6.1: 35–56.

——— (1999). "Bioregionalism—A Misplaced Project?" *Worldviews: Environment, Nature, Culture* 2.3: 215–37.

Brubaker, Rogers (1984). *The Limits of Rationality: An Essay on the Social and Moral Thought of Max Weber*. London: George Allen & Unwin.

Buchanen, Allen E. (1982). *Marx and Justice: The Radical Critique of Liberalism*. London: Methuen.

Burningham, Kate & Geoff Cooper (1999). "Being Constructive: Social Constructivism and the Environment." *Sociology* 33: 297–316.

Butler, Judith (1997). *The Psychic Life of Power: Theories in Subjection*. Stanford, Calif.: Stanford University Press.

Callicott, J. Baird (1984). "Non-Anthropocentric Value Theory and Environmental Ethics." *American Philosophical Quarterly* 21.4: 299–309.

——— (1985). "Intrinsic Value, Quantum Theory, and Environmental Ethics." *Environmental Ethics* 7: 257–75.

Callinicos, Alex (1989). *Against Postmodernism: A Marxist Critique*. Cambridge: Polity Press.

——— (1989a). *Making History: Agency, Structure and Change in Social Theory*. Cambridge: Polity Press.

Campbell, David and Michael J. Shapiro, eds. (1999). *Moral Spaces: Rethinking Ethics and World Politics*. Minneapolis: University of Minnesota Press.

Casey, Edward S. (1993). *Getting Back into Place*. Bloomington: Indiana University Press.

——— (1997). *The Fate of Place: A Philosophical History*. Berkeley: University of California Press.

Castoriadis, Cornelius (1992). "The Crisis of Marxism and Politics." *Society and Nature: The International Journal of Political Ecology* 2: 203–11.

Chakraverti, Satindranath (1978). "Praxis and Nature." In George F. Maclean (ed.), *Man and Nature*. Calcutta: Oxford University Press.

Chanter, Tina (1995). *Ethics of Eros: Irigaray's Rewriting of the Philosophers*. London: Routledge.

Cheney, Jim (1989). "Postmodern Environmental Ethics: Ethics as Bioregional Narrative." *Environmental Ethics* 11: 117–34.

——— (1989a). "The Neo-Stoicism of Radical Environmentalism." *Environmental Ethics* 11: 293–325.

——— (1994). "Nature/Theory/Difference: Ecofemininism and the Reconstruction of Environmental Ethics." In Karen Warren (ed.), *Ecological Feminism*. London: Routledge.

Churchill, Ward (1992). "False Promises: An Indigenist Examination of Marxist Theory and Practice." *Society and Nature: The International Journal of Political Ecology* 2: 212–33.

Clarke, John (1996). "Ajourd'hui l'ecologie? The French Take on Environmentalism." *Terra Nova: Nature and Culture* 1: 112–19.

Cohen, Gerry (1978). *Karl Marx's Theory of History*. Oxford: Clarendon Press.

Collier, Andrew (1994). "Value, Rationality and the Environment." *Radical Philosophy* 66: 3–9.

Comte, Auguste (1974). *The Positive Philosophy of Auguste Comte*, trans. Harriet Martineau. New York: AMS Press.

Conley, Verena Andermatt (1997). *Ecopolitics: The Environment in Poststructuralist Thought*. London: Routledge.

Cooke, Maeve (1999). "Questioning Autonomy: The Feminist Challenge and the Challenge for Feminism." In Richard Kearney & Mark Dooley (eds.), *Questioning Ethics: Contemporary Debates in Philosophy*. London: Routledge.

Coppe, Abiezer (1987). *Selected Writings*, ed. Andrew Hopton. London: Aporia Press.

Cronon, William (1989). *Changes in the Land: Indians, Colonists, and the Ecology of New England*. New York: Hill and Wang.

Cronon, William, ed. (1994). *Uncommon Ground: Toward Reinventing Nature*. New York: Norton.

Cross, F. L. (1966). *The Oxford Dictionary of the Christian Church*. London: Oxford University Press.

Darwin, Charles (1881). *The Formation of Vegetable Mould Through the Action of Worms with Observations on Their Habits*. London: John Murray.

Darwell, Stephen (1995). *The British Moralists and the Internal Ought 1640–1740*. Cambridge: Cambridge University Press.

Davidson, Joyce & Mick Smith (1999). "Wittgenstein and Irigaray: 'Philosophy in a Language (Game) of Difference'." *Hypatia: Journal of Feminist Philosophy* 14.2: 72–96.

Davies, Tony (1997). *Humanism*. London: Routledge.

Dawkins, Richard (1989). *The Selfish Gene*. Oxford: Oxford University Press.

Deleuze, Gilles (1988). *Spinoza: Practical Philosophy*. San Francisco: City Light Books.

Deleuze, Gilles & Felix Guattari (1987). *A Thousand Plateaus: Capitalism and Schizophrenia*. London: Athlone Press.

——— (1990). *Anti-Oedipus: Capitalism and Schizophrenia*. London: Athlone Press.

Derrida, Jacques (1978). *Speech and Phenomena*. Evanston, Ill.: Northwestern University Press.

Descola, Philippe & Gisli Palson, eds. (1996). *Nature and Society: An Anthropological Perspective*. London: Routledge.

De-Shalit, Avner (1996). *Why Posterity Matters: Environmental Policies and Future Generations*. London: Routledge.

Devall, Bill & George Sessions (1985). *Deep Ecology: Living as if Nature Mattered*. Salt Lake City: Peregrine Smith.

Docherty, Thomas, ed. (1993). *Postmodernism: A Reader*. Hemel Hempstead, U.K.: Harvester Wheatsheaf.

Dobson, Andrew (1990). *Green Political Thought*. London: HarperCollins.

Drysek, John S. (1990). "Green Reason: Communicative Ethics for the Biosphere." *Environmental Ethics* 12: 195–210.

Dunlap, Riley E. (1980). "Paradigmatic Change in Social Science: From Human Exemptions to an Ecological Paradigm." *American Behavioural Scientist* 24.1: 5–14.

Dunlap, Riley E. & W. E. Catton (1991). "Towards an Ecological Sociology." *The Annals of the International Institute of Sociology*, new series 3: 263–84.

Duerr, Hans Pater (1985). *Dreamtime: Concerning the Boundary between Wilderness and Civilisation*. Oxford: Blackwell.

Durkheim, Emile (1968). *The Elementary Forms of Religious Life*. London: George Allen & Unwin.

——— (1977). *Suicide: A Study in Sociology*. London: Routledge and Kegan Paul.

——— (1993). *Ethics and the Sociology of Morals*. Buffalo, N.Y.: Prometheus.

Eagleton, T. (1991). *Ideology: An Introduction*. London: Verso.

Eckersely, Robyn (1992). *Environmentalism and Political Theory: Toward an Ecocentric Approach*. London: University College London Press.

Eder, Klaus (1996). *The Social Construction of Nature*. London: Sage.

Engels, Frederick (1946). *Dialectics of Nature*. London: Lawrence and Wishart.

Esteva, Gustavo & Madhu Suri Prakash (1998). *Grassroots Postmodernism: Remaking the Soil of Cultures*. London: Zed.

Evans, Kate (1998). *Copse*. Chippenham, U.K.: Orange Dog Productions.

Featherstone, Mike (1988). "In Pursuit of the Postmodern: An Introduction." *Theory, Culture and Society*, 5.2–3: 195–215.

Featherstone, Mike, Scott Lash, & Roland Robertson, eds. (1995). *Global Modernities*. London: Sage.

Ferber, M & J. Nelson, eds. (1993). *Beyond Economic Man*. Chicago: University of Chicago Press.

Ferry, Luc (1992). *The New Ecological Order*. Chicago: University of Chicago Press.

Filoramo, Giovanni (1991). *A History of Gnosticism*. Oxford: Blackwell.

Findlay, J. N. (1970). *Axiological Ethics*. London: Macmillan.

Finn, Geraldine (1996). *Why Althusser Killed His Wife: Essays on Discourse and Violence*. Englewood Cliffs, N.J.: Humanities Press.

Flew, Anthony (1979). *A Dictionary of Philosophy*. London: Pan.

Foltz, Bruce V. (1995). *Inhabiting the Earth: Heidegger, Environmental Ethics, and the Metaphysics of Nature*. Englewood Cliffs, N.J.: Humanities Press.

Forest Alliance of British Columbia (1997). *Why Wood?* Vancouver.

Foster, John Bellamy (1992). "The Absolute General Law of Environmental Degradation under Capitalism." *Capitalism, Nature, Socialism: A Journal of Socialist Ecology* 11: 77–82.

Foucault, Michel (1980). *Power/Knowledge: Selected Interviews and Other Writings 1972–1977*. London: Harvester Wheatsheaf.

——— (1985). *The Use of Pleasure*. Harmondsworth, U.K.: Penguin.

——— (1986). *The Care of the Self*. Harmondsworth, U.K.: Penguin.

——— (1990). *The History of Sexuality: An Introduction*. Harmondsworth, U.K.: Penguin.

Frasz, Geoffrey B. (1993). "Environmental Virtue Ethics: A New Direction for Environmental Ethics." *Environmental Ethics* 15.3: 259–74.

Furrow, Dwight (1995). *Against Theory: Continental and Analytic Challenges in Moral Philosophy*. London: Routledge.

Gadamer, Hans-Georg (1998). *Truth and Method*. New York: Continuum.

Gadgil, Madhav & Ramachandra Guha (1995). *Ecology and Equity: The Use and Abuse of Nature in Contemporary India*. London: Routledge.

Gare, Arran E. (1995). *Postmodernism and the Environmental Crisis*. London: Routledge.

Gellner, Ernest (1968). *Words and Things*. Harmondsworth, U.K.: Penguin.

George, Paul et al. (1985). *Meares Island: Protecting a Natural Wilderness*. Tofino and Vancouver: Friends of Clayoquot Sound/Western Canada Wilderness Committee.

Gerrard, Nicci (1992). *The Observer Review*. June 16.

Giambelli, Rodolfo A. (1998). "The Coconut, the Body and the Human Being: Metaphors of Life and Growth in Nusa Pendida and Bali." In Laura Rival (ed.), *The Social Life of Trees: Anthropological Perspectives on Tree symbolism*. Oxford: Berg.

Gibbons, Maurice (1994). "The Clayoquot Papers." In *Clayoquot and Dissent*. Vancouver: Ronsdale.

Giddens, Anthony (1986). *Durkheim*. London: Fontana.

——— (1979). *Central Problems in Social Theory*. London: Macmillan.

——— (1996). *The Consequences of Modernity*. Cambridge: Polity Press.

Gilligan, Carol (1983). *In a Different Voice: Psychological Theory and Women's Development*. Cambridge, Mass.: Harvard University Press.

——— (1994). "Reply to My Critics." In Mary Jeanne Larabee (ed.), *An Ethic of Care: Feminist and Interdisciplinary Perspectives*. London: Routledge.

Ginsberg, Morris (1956). *On the Diversity of Morals*. London: Heinemann.

Golinski, Jan (1998). *Making Natural Knowledge: Constructivism and the History of Science*. Cambridge: Cambridge University Press.

Goodpaster, Kenneth (1978). "On Being Morally Considerable." *Journal of Philosophy* 75: 308–24.

Gordon, Avery F. (1997). *Ghostly Matters: Haunting and the Sociological Imagination*. Minneapolis: University of Minnesota Press.

Gorz, André (1983). *Farewell to the Working Class: An Essay on Post-Industrial Socialism*. London: Pluto Press.

——— (1985). *Paths to Paradise: On the Liberation from Work*. London: Pluto Press.

——— (1987). *Ecology as Politics*. London: Pluto Press.

Gottleib, Roger S. (1997). *The Ecological Community: Environmental Challenges for Philosophy, Politics and Morality*. London: Routledge.

Gould, Stephen Jay (1981). *The Mismeasure of Man*. New York: Norton.

Gowdy, John M. & Ada Ferreri Carbonell (1999). "Toward Consilience between Biology and Economics: The Contribution of Ecological Economics." *Ecological Economics* 29: 337–48.

Gregory, Derek (1997). "Lacan and Geography: The Production of Space Revisited." In Georges Benko & Ulf Strohmayer, (eds.), *Space and Social Theory: Interpreting Modernity and Postmodernity*. Oxford: Blackwell.

Grondin, Jean (1994). *Introduction to Philosophical Hermeneutics*. New Haven: Yale University Press.

Grove, Richard H. (1995). *Green Imperialism: Colonial Expansion, Tropical Island Edens and the Origins of Environmentalism, 1600–1860*. Cambridge: Cambridge University Press.

Grundmann, Reiner (1991). "The Ecological Challenge of Marxism." *New Left Review* 187: 103–20.

The Guardian (1996). "Trees Planted in Fast Lane at Road Protesters Party." July 15.

Habermas, Jürgen (1987). *The Philosophical Discourse of Modernity*. Cambridge: Polity Press.

——— (1987a). *The Theory of Communicative Action*, vol. 2: *The Critique of Functionalist Reason*. Cambridge: Polity Press.

——— (1990). *Moral Consciousness and Communicative Action*. Cambridge: Polity Press.

Hannigan, John (1995). *Environmental Sociology: A Social Constructivist Perspective*. London: Routledge.

Hardin, Garrett (1968). "The Tragedy of the Commons." *Science* 162: 1243–48.

Harding, Sandra (1991). *Whose Science? Whose Knowledge? Thinking from Women's Lives*. Milton Keynes, U.K.: Open University Press.

Hare, R. M. (1952). *The Language of Morals*. Oxford: Oxford University Press.

Harr, Michel (1993). *The Song of the Earth: Heidegger and the Grounds of the History of Being*. Bloomington: Indiana University Press.

Harris, Karsten (1997). *The Ethical Function of Architecture*. Cambridge, Mass.: MIT Press.

Harvey, David (1996). *Justice, Nature and the Geography of Difference*. Oxford: Blackwell.

Harvey, Paul H. & Paul J. Greenwood (1978). "Anti-predator Defence Strategies: Some Evolutionary Problems." In John R. Krebs & N. B. Davies (eds.), *Behavioural Ecology: An Evolutionary Approach*. Oxford: Blackwell.

Haywood, Tim (1992). "Ecology and Human Emancipation." *Radical Philosophy*, 62: 3–13.

——— (1994). "The Meaning of Political Ecology." *Radical Philosophy*, 66: 11–20.

Heidegger, Martin (1993). "The Question Concerning Technology." In *Basic Writings*, ed. David Farrell Krell. London: Routledge.

Hekman, Susan J. (1983). *Max Weber and Contemporary Social Theory*. Oxford: Martin Robertson.

——— (1995). *Moral Voices, Moral Selves: Carol Gilligan and Feminist Moral Theory*. Cambridge: Polity Press.

Held, David (1980). *Introduction to Critical Theory: Horkheimer to Habermas*. London: Hutchinson.

Held, Virginia (1995). *Justice and Care: Essential Readings in Feminist Ethics*. Boulder, Colo.: Westview Press.

Hepworth, James R. & Gregory McNamee, eds. (1996). *Resist Much Obey Little: Remembering Ed Abbey*. San Francisco: Sierra Club Books.

Hill, Christopher (1991). *The World Turned Upside Down: Radical Ideas during the English Revolution*. Harmondsworth, U.K.: Penguin.

——— (1996). *Liberty against the Law: Some Seventeenth Century Controversies*. Harmondsworth, U.K.: Penguin.

Hollis, Martin & Edward J. Nell (1975). *Rational Economic Man: A Philosophical Critique of Neoclassical Economics*. Cambridge: Cambridge University Press.

Horkheimer, Max (1974). *Eclipse of Reason*. New York: Seabury Press.

Horn, Henry S. (1981). "Succession." In Robert M. May (ed.), *Theoretical Ecology: Principles and Applications*. Oxford: Blackwell Scientific Publications.

Howell, Signe (1977). *The Ethnography of Moralities*. London: Routledge.

Hoy, David (1990). "Derrida." In Quentin Skinner (ed.), *The Return of Grand Theory in the Human Sciences*. Cambridge: Cambridge University Press.

Irigaray, Luce (1985). *This Sex Which Is Not One*. Ithaca, N.Y.: Cornell University Press.

——— (1992). *Elemental Passions*. London: Athlone Press.

——— (1993). *An Ethics of Sexual Difference*. London: Athlone Press.

——— (1993a). *je, tu, nous: Toward a Culture of Difference*. London: Routledge.

Irwin, Alan (1995). *Citizen Science: A Study of People, Expertise and Sustainable Development*. London: Routledge.

Ivanson, Duncan (1998). "The Disciplinary Moment: Foucault, Law and the Rejuvenation of Rights." In Jeremy Moss (ed.), *The Later Foucault*. London: Sage.

Jay, Martin (1973). *The Dialectical Imagination: A History of the Frankfurt School and the Institute of Social Research 1923–1950*. London: Heinemann.

Jameson, Frederick (1991). *Postmodernism or the Cultural Logic of Late Capitalism*. London: Verso.

Jenkins, Richard (1992). *Pierre Bourdieu*. London: Routledge.

Johnson, Lawrence (1991). *A Morally Deep World: An Essay on Moral Significance and Environmental Ethics*. Cambridge: Cambridge University Press.

Kahlberg, Stephen (1980). "Max Weber's Types of Rationality: Cornerstones for the Analysis of Rationalisation Processes in History." *American Journal of Sociology*, 85.5: 1145–79.

Kamenka, Eugene (1983). *The Portable Karl Marx*. Harmondsworth, U.K.: Viking Penguin.

Käsler, Dirk (1988). *Max Weber: An Introduction to His Life and Work*. Cambridge: Polity Press.

Keith, Michael & Steve Pile, eds. (1993). *Place and the Politics of Identity*. London: Routledge.

Kellert, Stephen R. & Edmund O. Wilson (1993). *The Biophilia Hypothesis*. Washington, D.C.: Island Press.

Kern, Stephen (1983). *The Culture of Time and Space 1880–1918*. Cambridge, Mass.: Harvard University Press.

Kohlberg, Lawrence (1981). *The Philosophy of Moral Development*. San Francisco: Harper & Row.

Kilminster, Richard (1989). "Sociology and the Professional Culture of Philosophers." In Hans Haferkamp (ed.), *Social Structure and Culture*. New York: de Gruyter.

Kirby, Vicki (1997). *Telling Flesh: The Substance of the Corporeal*. London: Routledge.

Korsgaard, Christine M. (1996). *Sources of Normativity*. Cambridge: Cambridge University Press.

Krebs, Charles J. (1978). *Ecology: The Experimental Analysis of Distribution and Abundance*, 2nd ed. New York: Harper & Row.

Lacan, Jaques (1992). *Écrits: A Selection*. London: Routledge.

Laclau, Ernesto & Chantal Mouffe (1987). *Hegemony and Socialist Strategy: Towards a Radical Democratic Politics*. London: Verso.

Lakatos, Imre & Alan Musgrave, eds. (1972). *Criticism and the Growth of Knowledge*. Cambridge: Cambridge University Press.

Landry, Donna & Gerald MacLean (1993). *Materialist Feminisms*. Oxford: Blackwell.

Larrabee, Mary Jeanne, ed. (1993). *An Ethic of Care: Feminist and Interdisciplinary Perspectives*. London: Routledge.

Lash, Scott & John Urry (1994). *Economies of Signs and Space*. London: Sage.

Laska, S. B. (1993). "Environmental Sociology and the State of the Discipline." *Social Forces* 72.1: 1–17.

Latour, Bruno & Steven Woolgar (1986). *Laboratory Life: The Construction of Scientific Facts*. Princeton, N.J.: Princeton University Press.

Le Corbusier (1971). *The City of Tomorrow and Its Planning*. London: Architectural Press.

Lee, Martha F. (1995). *Earth First!: Environmental Apocalypse*. Syracuse, N.Y.: Syracuse University Press.

Lefebvre, Henri (1994). *The Production of Space*. Oxford: Blackwell.

——— (1995). *Introduction to Modernity. Twelve Preludes, September 1959 – May 1961*. London: Verso.

Leff, Enrique (1992). "A Second Contradiction of Capitalism? Notes for the Environmental Transformation of Historical Materialism." *Capitalism, Nature, Socialism: A Journal of Socialist Ecology* 12: 109–16.

——— (1995). *Green Production: Toward an Environmental Rationality*. New York: Guilford Press.

Leopold, Aldo (1949). *A Sand County Almanac*. New York: Oxford University Press.

Levinas, Emmanuel (1991). *Totality and Infinity*. London: Kluwyer Academic Publicatıons.

Lévi-Strauss, Claude (1968). *Structural Anthropology*. London: Allen Lane.

Lloyd, Genevieve (1984). *The Man of Reason: Male and Female in Western Philosophy*. London: Methuen.

Long, A. A. (1974). *Hellenistic Philosophy*. London: Duckworth.

Lovejoy, Arthur O. (1964). *The Great Chain of Being: A Study of the History of an Idea*. Cambridge, Mass.: Harvard University Press.

Lovelock J. E. (1979). *Gaia: A New Look at the Earth*. Oxford. Oxford University Press.

——— (1988). *The Ages of Gaia*. Oxford. Oxford University Press.

Lukács, Georg (1978). *The Ontology of Social Being*, vol. 3: *Labour*. London: Merlin Press.

Luhmann, Niklas (1989). *Ecological Communication*. Chicago: University of Chicago Press.

Lukes, Steven (1985). *Marxism and Morality*. Oxford: Clarendon.

——— (1988). *Emile Durkheim*. Harmondsworth, U.K.: Penguin.

Lukes, Steven & Andrew Scott (1984). *Durkheim and the Law*. Oxford: Blackwell.

Lynch, Michael (1998). "Towards a Constructivist Genealogy of Social Constructivism." In Irving Velody & Robin Williams (eds.), *The Politics of Social Constructionism*. London: Sage.

Lyotard, Jean-François (1991). *The Postmodern Condition: A Report on Knowledge*. Manchester: Manchester University Press.

——— (1997). *Postmodern Fables*. Minneapolis: University of Minnesota Press.

Macauley, David (1992). "Out of Place and Outer Space: Hannah Arendt on Earth Alienation: A Historical and Critical Perspective." *Capitalism, Nature, Socialism* 3.4: 19–45.

MacIntyre, Alisdair (1966). *A Short History of Ethics*. London: Routledge & Kegan Paul.

——— (1981). "A Crisis in Moral Philosophy." In D. Callghan & H. T. Englehardt Jr. (eds.), *The Roots of Ethics: Science, Religion and Values*. New York: Plenum Press.

Mackie, J. L. (1978). *Ethics: Inventing Right and Wrong*. Harmondsworth, U.K.: Penguin.

Macnaghten, Phil & John Urry (1998). *Contested Natures*. London: Sage.

MacPherson, C. B. (1979). *The Political Theory of Possessive Individualism: Hobbes to Locke*. Oxford: Oxford University Press.

Malpas, J. (1994). "A Taste of Madeleine: Notes Towards a Philosophy of Place." *International Philosophical Quarterly* 34: 433–51.

Marcuse, Herbert (1968). *Negations: Essays in Critical Theory*. Harmondsworth, U.K.: Penguin.

——— (1991). *One-Dimensional Man: Studies in the Ideology of Advanced Industrial Society*. London: Routledge.

——— (1992). "Ecology and the Critique of Modern Society." *Capitalism, Nature, Socialism* 3.3: 29–38.

——— (1998). *Technology, War and Fascism: Collected Papers, Vol. 1*. London: Routledge.

Marvell, Andrew (1993). *The Complete Poems*. London: Everyman.

Marx, Karl (1970). "Critique of the Gotha Programme." In Ernst Fischer (ed.), *Marx in His Own Words*. Harmondsworth, U.K.: Penguin.

——— (1973). *Grundrisse*. Harmondsworth, U.K.: Penguin.

——— (1977). *Selected Writings*, ed. David McLellan. Oxford: Oxford University Press.

——— (1990). *Capital*, vol. 1. Harmondsworth U.K.: Penguin.

Marx, Karl & Frederick Engels (1977). *The Manifesto of the Communist Party*. Peking.

Mathews, Freya (1993). *The Ecological Self*. London: Routledge.

Mauss, Marcel & Henri Beuchat (1979). *Seasonal Variations of the Eskimo: A Study in Social Morphology*. London: Routledge and Kegan Paul.

McCarthy, Thomas (1984). *The Critical Theory of Jurgen Habermas*. Cambridge: Polity Press.

McKay, George (1998). *DIY Culture: Party and Protest in Nineties Britain*. London: Verso.

McAllister, Donald (1980). *Evaluation in Environmental Planning*. Cambridge, Mass.: M.I.T. Press.

McGinnis, Michael Vincent (1999). "A Rehearsal to Bioregionalism." In Michael Vincent McGinnis (ed.), *Bioregionalism*. London: Routledge.

McKibben, Bill (1990). *The End of Nature*. Harmondsworth, U.K.: Penguin.

McLellan, David (1971). *Marx's Grundrisse*, St. Albans, U.K.: Paladin.

Merchant, Carolyn (1990). *The Death of Nature: Women, Ecology and the Scientific Revolution*. San Francisco: Harper & Row.

Merchant, Carolyn (1996). *Earthcare: Women and the Environment*. London: Routledge.

Merrick (1997). *The Battle for the Trees*. Leeds, U.K.: Godhaven.

M'Gonigle, R. Michael (1999). "Ecological Economics and Political Ecology: Towards a Necessary Synthesis." *Ecological Economics* 28: 11–26.

Mill, John Stuart (1885). *Three Essays on Religion*. London: Longmans Green & Co.

Mill, John Stuart & Jeremy Bentham (1987). *Utilitarianism and Other Essays*. Harmondsworth, U.K.: Penguin.

Milton, Kay (1996). *Environmentalism and Cultural Theory: Exploring the Role of Anthropology in Cultural Discourse*. London: Routledge.

Monbiot, George (1994). "Defiant Culture." *The Guardian*, December 7.

Morgan, Loýs (1994). "Clayoquot: Recovering from Cultural Rape." In *Clayoquot and Dissent*. Vancouver: Ronsdale.

Muir, John (1988). *My First Summer in the Sierra*. Edinburgh: Cannongate Classics.

Mumford, Lewis (1947). *Technics and Civilisation*. London: Routledge.

Murdoch, Iris (1970). *The Sovereignty of Good*. London: Routledge & Kegan Paul.

Myerson, George & Yvonne Rydin (1996). *The Language of the Environment: A New Rhetoric*. London: University College London Press.

Naess, Arne (1972). "The Shallow and the Deep Ecology Movement." *Inquiry* 16: 95–100.

——— (1979). "Self-Realisation in Mixed Communities of Humans, Bears, Sheep and Wolves." *Inquiry*, 22: 231–41.

Neilson, Kai (1991). "Does a Marxian Critical Theory of Society Need a Moral Theory?" *Radical Philosophy* 59: 21–26.

Newby, Howard (1991). "One World/Two Cultures: Sociology and the Environment." *Network* 50.

Newton-Smith, W. H. (1981). *The Rationality of Science*. London: Routledge & Kegan Paul.

Nietzsche, Freidrich (1966). "On Truth and Lie in an Extra Moral Sense." *The Portable Nietzsche*. London: Macmillan.

Norris, Christoper (1990). *What's Wrong with Postmodernism?* London: Harverster Wheatsheaf.

Nye, Andrea (1988). *Feminist Theory and the Philosophies of Man*. London: Croom Helm.

O'Connor, James (1991). "On the Two Contradictions of Capitalism." *Capitalism, Nature, Socialism: A Journal of Socialist Ecology* 2.3.

Oelschlaeger, Max (1991). *The Idea of Wilderness*. New Haven: Yale University Press.

———, ed. (1995). *Postmodernism and the Environment*. Albany: State University of New York Press.

O'Neill, John (1993). *Ecology, Policy, Politics: Human Well-Being and the Natural World*. London: Routledge.

Parfit, Derek (1984). *Reasons and Persons*. Oxford: Clarendon Press.

Parsons, Talcott (1970). *The Social System*. London: Routledge & Kegan Paul.

Pearce, David, Anil Markandya, & Edward B. Barbier (1989). *Blueprint for a Green Economy*. London: Earthscan.

——— (1995). *Blueprint 4: Capturing Environmental Value*. London: Earthscan.

Peperzak, Adriaan (1993). *To the Other: An Introduction to the Philosophy of Emmanuel Levinas*. West Lafayette, Ind.: Purdue University Press.

Peperzak, Adriaan T., Simon Critchly, & Robert Bernasconi, eds. (1996). *Emmanuel Levinas: Basic Philosophical Writings*. Bloomington and Indianapolis: Indiana University Press.

Pepper, David (1993). *Eco-socialism: From Deep Ecology to Social Justice*. London: Routledge.

Petagorskey, David W. (1940). *Left-Wing Democracy in the English Civil War*. London: Gollancz.

Pile, Steve & Michael Keith, eds. (1997). *Geographies of Resistance*. London: Routledge.

Plumwood, Val (1993). *Feminism and the Mastery of Nature*. London: Routledge.

Poane, Sergio (1997). "Cutting of Ancient Forests on Vancouver Island about to Increase." Tofino, British Columbia: Friends of Clayquot Sound.

Ponting, Clive (1991). *A Green History of the World*. London: Sinclair-Stevenson.

Proctor, James D. & David M. Smith, eds. (1999). *Geography and Ethics: Journeys in a Moral Terrain*. London: Routledge.

Quinton, Anthony (1995). "Analytic Philosophy." In Ted Honderich (ed.), *The Oxford Companion to Philosophy*. Oxford: Oxford University Press.

Raven, Charlotte (1997). *The Guardian*, April 22.

Reaves, Dixie Watts, Randall A. Kramer, & Thomas P. Holmes (1999). "Does Question Format Matter? Valuing Endangered Species." *Environmental and Resource Economics* 14: 365–83.

Ree, Jonathan (1993). "English Philosophy in the Fifties." *Radical Philosophy* 65: 3–21.

Ricoeur, Paul (1986). *Lectures on Ideology and Utopia*. New York: Columbia University Press.

Rival, Laura, ed. (1998). *The Social Life of Trees: Anthropological Perspectives on Tree Symbolism*. Oxford: Berg.

Robbins, Derek (1991). *The Work of Pierre Bourdieu: Recognising Society*. Milton Keynes, U.K.: Open University Press.

Robins, Kevin (1995). "Cyberspace and the World We Live In." In Mike Featherstone & Roger Burrows (eds.), *Cyberspace, Cyberbodies, Cyberpunk: Cultures of Technological Embodiment*. London: Sage.

Rodman, John (1983). "Four Forms of Ecological Consciousness." In Donald Scherer & Thomas Attig (eds.), *Ethics and the Environment*. Englewood Cliffs, N.J.: Prentice Hall.

Rolston, Holmes (1997). "Nature for Real: Is Nature a Social Construction?" In T. D. J. Chappell (ed.), *The Philosophy of the Environment*, Edinburgh: Edinburgh University Press.

Rorty, Richard (1988). *Philosophy and the Mirror of Nature*. Oxford: Blackwell.

Rose, Gillian (1993). *Feminism and Geography: The Limits of Geographical Knowledge*. Cambridge: Polity Press.

Rose, Gillian (1996). "As if the Mirrors Had Bled: Masculine Dwelling, Masculinist Theory and Feminist Masquerade." In Nancy Duncan (ed.), *Body Space: Destabalizing Geographies of Gender and Sexuality*. London: Routledge.

Ross, Andrew (1994). *The Chicago Gangster Theory of Life: Nature's Debt to Society*. London: Verso.

Rousseau, Jean-Jacques (1979). *The Reveries of a Solitary Walker*. Harmondsworth, U.K.: Penguin.

Rudder-Baker, Lynn (1984). "On the Very Idea of a Form of Life." *Inquiry* 27: 277–89.

Ryle, Martin (1988). *Ecology and Socialism*. London: Century Hutchinson.

Sabine, George H. (1963). *A History of Political Theory*. London: Harrap.

Sagoff, Mark (1989). *The Economy of the Earth: Philosophy, Law, and the Environment*. Cambridge: Cambridge University Press.

Sale, Kirkpatrick (1991). *The Conquest of Paradise: Christopher Columbus and the Columbian Legacy*. London: Hodder & Stoughton.

Sale, Kirkpatrick (1991). *Dwellers in the Land: The Bioregional Vision*. Philadelphia: New Society.

Salleh, Ariel (1997). *Ecofeminism as Politics: Nature, Marx and the Postmodern*. London: Zed Books.

Sargent, Lydia, ed. (1981). *The Unhappy Marriage of Marxism and Feminism: A Debate on Class and Patriarchy*. London: Pluto Press.

Schama, Simon (1996). *Landscape and Memory*. London: Fontana.

Schivelsbusch, Wolfgang (1986). *The Railway Journey: The Industrialisation of Time and Space in the Nineteenth Century*. Leamington Spa, U.K.: Berg.

Schmidt, Alfred (1971). *The Concept of Nature in Marx*. London: New Left Books.

Schutz, Alfred (1967). *The Phenomenology of the Social World*. Evanston, Ill.: Northwestern University Press.

Seidler, Victor J. (1994). *Recovering the Self: Morality and Social Theory*. London: Routledge.

Seidman, Stephen (1992). "Postmodern Social Theory as Narrative with Moral Intent." In Stephen Seidman & D. G. Wagner (eds.), *Postmodern Social Theory*. Oxford: Blackwell.

Seel, Ben (1998). "Strategies of Resistance at the Pollock Free State Road Protest Camp." *Environmental Politics* 6.4

Sen, Amartya (1977). "Rational Fools: A Critique of the Behavioural Foundations of Economic Theory." *Philosophy and Public Affairs* 6: 317–44.

Serres, Michel (1998). *The Natural Contract.* Ann Arbour: University of Michigan Press.

Sessions, George (1995). "Postmodernism, Environmental Justice, and the Demise of the Ecology Movement?" *Wild Duck Review* 5.

Seymour, John & Herbert Giradet (1990). *Far From Paradise: The Story of Human Impact on the Environment.* London: Greenprint.

Shakespeare, William (1971). *The Tempest.* London: The Folio Society.

Shiva, Vandana (1994). *Staying Alive: Women, Ecology and Development.* London: Zed Books.

Short, John Rennie (1991). *Imagined Country: Society, Culture and Environment.* London: Routledge.

Sibley, David (1995). *Geographies of Exclusion.* London: Routledge.

Silber, Ilana Friedrich (1995). "Space, Fields, Boundaries: The Rise of Spatial Metaphors in Contemporary Sociological Theory." *Social Research* 62.2: 323–55.

Simmonds, I. G. (1991). *Changing the Face of the Earth: Culture, Environment, History.* Oxford: Blackwell.

——— (1994). *Environmental History: A Concise Introduction.* Oxford: Blackwell.

Singer, Peter (1981). *The Expanding Circle, Ethics and Sociobiology.* Oxford: Clarendon Press.

——— (1990). *Animal Liberation.* London: Thorsons.

Skinner, Quentin (1990). *The Return of Grand Theory in the Human Sciences.* Cambridge: Cambridge University Press.

Smart, J. J. C. & Bernard Williams (1990). *Utilitarianism For and Against.* Cambridge: Cambridge University Press.

Smith, Dorothy (1987). *The Everyday World as Problematic: A Feminist Sociology.* Boston: Northeastern University Press.

Smith, Mick. (1991). "Letting in the Jungle." *Journal of Applied Philosophy* 8.2: 145–54.

——— (1992). Review of Lawrence E. Johnson's "A Morally Deep World." *Environmental Values* 1.1: 88–90.

——— (1993). "Cheney and the Myth of Postmodernism." *Environmental Ethics* 15: 3–17.

——— (1995). "A Green Thought in a Green Shade: Against the Rationalisation of Environmental Values." In Yvonne Guerrier, Nicholas Alexander,

Jonathan Chase, & Martin O'Brian (eds.), *Values and the Environment: A Social Science Perspective*. Chichester, U.K.: John Wiley.

——— (1997). "The Enclosure of the Ethical Commons: Radical Environmentalism as an Ethics of Place." *Environmental Ethics* 18.4: 339–53.

——— (1997a). "What's Natural? The Socio-Political (De)construction of Nature." *Environmental Politics* 6.2: 164–68.

——— (1998). "The Ethical Architecture of the 'Open Road.'" *Worldviews: Environment, Religion, Culture* 2: 185–99.

——— (1998a). "Vestigial Philosophy: Academia and the Institutionalisation of Thought." *Abertay Sociology Papers*, no. 1.

Smith, Neil & Cindi Katz (1993). "Grounding Metaphor: Towards a Spatialized Politics." In Michael Keith & Steve Pile (eds.), *Place and the Politics of Identity*. London: Routledge.

Snyder, Gary (1984). *Good, Wild, Sacred*. Hereford, U.K.: Five Seasons Press.

Söderbaum, Peter (1999). "Values, Ideology and Politics in Ecological Economics." *Ecological Economics* 28: 161–70.

Soja, Edward (1990). *Postmodern Geographies: The Re-assertion of Space in Critical Social Theory*. London: Verso.

——— (1996). *Thirdspace: Journeys to Los Angeles and Other Real and Imagined Places*. Oxford: Blackwell.

Soper, Kate (1986). *Humanism and Anti-humanism*. London: Hutchinson.

——— (1987). "Marxism and Morality." *New Left Review* 163: 101–13.

Soulé, Michael & Gary Lease, eds. (1995). *Reinventing Nature? Responses to Postmodern Deconstruction*. Washington, D.C.: Island Press.

Spector, M. & J. I. Kitsuse (1967). *Constructing Social Problems*. Menlo Park, Calif.: Cummings.

Spinoza, Baruch (1982). *The Ethics and Selected Letters*. Indianapolis: Heckett Publishing.

Stone, Christopher (1974). *Should Trees Have Standing?* Los Angeles: Kaufman.

Sturrock, John, ed. (1979). *Structuralism and Since*. Oxford: Oxford University Press.

Sunday Sport (1997). May 25.

Tall, Deborah (1996). "Dwelling: Making Peace with Space and Place." In William Vitek and Wes Jackson (eds.), *Rooted in the Land: Essays on Community and Place*. New Haven: Yale University Press.

Tar, Zoltan (1977). *The Frankfurt School: The Critical Theory of Max Horkheimer and Theodor W. Adorno*. New York: John Wiley.

Tate, W. E. (1967). *The Enclosure Movement*. New York: Walker & Co.

Taylor, Bron (1997). "Earth First! Fight Back." *Terra Nova* 2.2: 26–41.

Taylor, Paul W. (1986). *Respect for Nature: A Theory of Environmental Ethics*. Princeton, N.J.: Princeton University Press.

Tester, Keith (1991). *Animals and Society: The Humanity of Animal Rights*. London: Routledge.

——— (1997). *Moral Culture*. London: Sage.

Thomas, Keith (1984). *Man and the Natural World: Changing Attitudes in England 1500–1800*. Harmondsworth, U.K.: Penguin.

Thomashow, Michael (1999). "Towards a Cosmopolitan Bioregionalism." In Michael Vincent McGinnis (ed.), *Bioregionalism*. London: Routledge.

Thompson, E. P. (1978). *The Poverty of Theory*. London: Merlin.

——— (1994). *Witness against the Beast: William Blake and the Moral Law*. Cambridge: Cambridge University Press.

Tobias, Michael, ed. (1984). *Deep Ecology*. San Marcos, Calif.: Avant.

Toledo, Victor (1992). "The Ecological Crisis: A Second Contradiction of Capitalism." *Capitalism, Nature, Socialism: A Journal of Socialist Ecology* 3.3: 84–88.

Touraine, Alain (1995). *Critique of Modernity*. Oxford: Blackwell.

Tuck, Richard (1977). *Natural Rights Theories: Their Origins and Development*. Cambridge: Cambridge University Press.

Turner, Victor (1985). *Rediscovering America: John Muir in His Time and Ours*. San Francisco: Sierra Club Books.

Uchiyamada, Yasushi (1998). "'The Grove is our Temple': Contested Representations of *Kaavu* in Kerala, South India." In Laura Rival (ed.), *The Social Life of Trees: Anthropological Perspectives on Tree Symbolism*. Oxford: Berg.

Vidal, John (1996). *The Guardian*, March 13.

——— (1998). *The Guardian*, April 1.

Virilio, Paul (1997). *Open Sky*. London: Verso.

Vogel, Lise (1983). *Marxism and the Oppression of Women: Towards a Unitary Theory*. London: Pluto Press.

Vogel, Steven (1996). *Against Nature: The Concept of Nature in Critical Theory*. Albany: State University of New York Press.

Vogel, Steven (1997). "Habermas and the Ethics of Nature." In Roger S. Gottleib, *The Ecological Community: Environmental Challenges for Philosophy, Politics and Morality*. London: Routledge.

Wainwright, Martin (1997). *The Guardian*, May 21.

Waldron, Jeremy (1987). *Nonsense upon Stilts: Bentham, Burke and Marx on the Rights of Man*. London: Methuen.

Wall, Derek (1999). *Earth First! and the Radical Anti-Roads Movement: Radical Environmentalism and Comparative Social Movements*. London: Routledge.

Wallace, Ruth A. & Alison Wolf (1995). *Contemporary Sociological Theory: Continuing the Classical Tradition*, 4th ed. Englewood Cliffs, N.J.: Prentice Hall.

Walzer, Michael (1995). "The Politics of Michel Foucault." In David Hoy (ed.), *Foucault: A Critical Reader*. Oxford: Basil Blackwell.

Ward, David (1997). *The Guardian*, May 21.

Weber, Max (1964). *The Theory of Social and Economic Organisation*. New York: Free Press.

——— (1964). *Max Weber: Essays in Sociology*. London: Routledge & Kegan Paul.

Weiner, Norbert (1948). *Cybernetics: or Control and Communication in the Animal and the Machine*. New York: John Wiley.

Western Forest Products Limited (1995–96). *Visitors' Guide: Welcome to Northern Vancouver Island*.

Whatmore, Sarah (1997). "Dissecting the Autonomous Self: Hybrid Cartographies for a Relational Ethics." *Environment and Planning D: Society and Space* 15: 37–53.

White, Stephen K. (1990). *The Recent Work of Jurgen Habermas: Reason, Justice & Modernity*. Cambridge: Cambridge University Press.

Whitebook, Joel (1979). "The Problem of Nature in Habermas." *Telos* 40: 41–69.

Whitford, Margaret (1991). *Luce Irigaray Philosophy in the Feminine*. London: Routledge.

Whimster, Sam & Scott Lash, eds. (1987). *Max Weber, Rationality and Modernity*. London: Allen & Unwin.

Wiggerhaus, Rolf (1986). *The Frankfurt School: Its History, Theories and Political Significance*. Cambridge: Polity Press.

Wilderness Committee (1997). "Wild Salmon—Majestic Grizzlies—Ancient Rainforests." *Wilderness Committee Educational Report* 16.4.

Williams, Raymond (1989). *Problems in Materialism and Culture: Selected Essays*. London: Verso.

Wilson, Alexander (1992). *The Culture of Nature: North American Landscape from Disney to the Exxon Valdez*. Cambridge, Mass.: Blackwell.

Wilson, E. O. (1998). *Consilience: The Unity of Knoweldge*. London: Little, Brown and Company.

Wittgenstein, Ludwig (1988). *Philosophical Investigations*. Oxford: Blackwell.

Wolch, Jennifer & Jody Emel, eds. (1998). *Animal Geographies: Place, Politics and Identity in the Nature-Culture Borderlands*. London: Verso.

Yelling, J. A. (1977). *Common Field and Enclosure in England 1450–1850*. London: Macmillan.

Young, Michael, (1978). "Althusser's Marxism and British Social Science." *Radical Science* 6/7: 129–34.

Zerzan, John (1994). *Future Primitive and Other Essays*. New York: Autonomedia.

Zimmerman, Michael E. (1990). *Heidegger's Confrontation with Modernity: Technology, Politics, Art*. Bloomington: Indiana University Press.

Zimmerman, Michael E. (1994). *Contesting Earth's Future: Radical Ecology and Postmodernity*. Berkeley: University of California Press.

Index